W9-BOA-742

The Complete
What's Your Poo Telling You?

The Complete
What's Your Poo Telling You?

Josh Richman and Anish Sheth, M.D.

**Illustrations by
Peter Arkle and Matt Johnstone**

CHRONICLE BOOKS
SAN FRANCISCO

Text copyright © 2018 by Josh Richman and
Dr. Anish Sheth
Illustrations copyright © 2007 and 2009 by
Peter Arkle and Matt Johnstone

All rights reserved. No part of this book may
be reproduced in any form without written
permission of the publisher.

Library of Congress Cataloging-in-
Publication data is available.

ISBN 978-1-4521-7007-7

Manufactured in China.

10 9 8 7 6 5 4 3 2 1

Chronicle Books LLC
680 Second Street
San Francisco, CA 94131
www.chroniclebooks.com

Contents

Introduction

The Deuce is loose! A decade after it first dropped, we bring you *The Complete What's Your Poo Telling You?* This is no simple case of déjà poo; this glorious release brings together the best poo knowledge from *What's Your Poo Telling You?*, *What's My Pee Telling Me?*, and ten years of daily calendar entries all together in one place. Poo remains the star of the show, but not to worry, there is plenty of pee and flatulence to go around.

Amidst our ever-changing and divided world, we are reminded that everyone poops. Whether you visit the john once a day (yay!) or once a week (yikes!), we can agree that the experiences of eliminating solid, liquid, and gaseous waste are universal.

We are grateful (and, yes, relieved) that so many of you have taken the plunge into this fascinating world. By compacting all the best nuggets together, *The Complete What's Your Poo Telling You?* aims to keep things in motion, making trips to the loo more enjoyable and informed by combining tongue-in-cheek humor with real medical information.

The new edition contains all the classics, from floaters to sinkers, clean sweeps to N.E.W.P (Never-Ending Wipe Poo). Discover the joy of Poo-phoria and the agony of the Chinese

Star. Poo's myriad colors, shapes, consistencies, and aromas are all broken down so you can make sense of your digestive health.

This new double digest builds upon the first movement and may inspire you to spend extra time perched on your throne; just don't dally too long (turns out reading on the toilet for long periods of time can cause hemorrhoids!).

And remember, as always, look before you flush!

— Josh Richman and Anish Sheth, M.D.

CHAPTER 1:
Poo 101

Understanding poo requires a knowledge of anatomy, biology, chemistry, and physics, not to mention history and art.

A Picture Is Worth a Thousand Words . . .

The Bristol Stool Scale is a tool used by gastroenterologists to help standardize the discourse on poo. Utilizing pictures and descriptions such as "like a sausage but with cracks," the scale breaks down poo into seven types. British physicians first published this poo pictorial in 1990 and found that the appearance of poo correlated with the amount of time it spends in the colon; the harder and more lumpy the stool, the slower the travel time.

BRISTOL STOOL CHART

Type 1: Separate hard lumps, like nuts (hard to pass)
Type 2: Sausage-shaped but lumpy
Type 3: Like a sausage but with cracks on its surface
Type 4: Like a sausage or snake, smooth and soft
Type 5: Soft blobs with clear-cut edges (passed easily)
Type 6: Fluffy pieces with ragged edges, a mushy stool
Type 7: Watery, no solid pieces (entirely liquid)

DOO YOU KNOW?

WHAT'S YOUR POO MADE OF?

- Ten parts water
- One part bacteria (dead and alive)
- One part indigestible fiber
- One part mixture of fat, protein, dead cells, and mucus

The Ritual Poo

Synonyms: *Rite of Passage, Palm Poo, Calendar Crap*

This poo occurs at the same time each day. Due to the clockwork nature of this poo, you can prepare by having a newspaper ready and your favorite bathroom location scoped out ahead of time. The regularity of these poos may provide comfort in an otherwise unpredictable world.

Dr. Stool says: "Just as the sun will rise ... " The early-morning ritual poo can be a great start to the day. Some experience this poo only after consuming their ritual morning coffee (caffeine works as a laxative by increasing the colon's contractions). For others, the poo trigger is fired after each meal. This "postprandial" poo is caused by the gastrocolic reflex, where distension of the stomach by meal contents causes a reflex stimulation of the intestines, moving stool into the rectum and giving us the urge to defecate. (Translation: Out with the old, in with the new.) Need proof of this biological truth? Check out the bathroom stalls at work about a half-hour after lunch.

The Clean Sweep

Synonyms: *Wipeless Poo, The Perfect Wipe, Mr. Clean*
Antonyms: *The Wet Poo, Swamp Ass, Mudslide*

On rare and special occasions, you engage in the entire stooling process from engagement to deployment and note, in the cleanup phase, that amazingly there is no poo residue on the toilet paper! Despite coming up clean on the first wipe, some skeptical souls wipe one more time just to make sure there isn't a poo illusion at play. Some experts consider the "wipeless" poo to be the pinnacle of poo performance. You may depart the restroom feeling extra clean after this type of poo.

This is in direct contrast to other times when you use half the roll of toilet paper and feel as if you haven't made any progress. What's worse is that those moist, multi-wipe poos seem to occur at the most inopportune times—e.g., in a crowded public restroom where you only have a few inches of cheap, single-ply, essentially transparent toilet paper. In these instances, wiping can be so unsuccessful that you end up putting toilet paper between your butt cheeks to avoid painting your underwear with skid marks. These situations, juxtaposed to Clean Sweeps, remind us that although there is little discernible difference during the exit, the cleanup can have dramatic variation.

Dr. Stool says: In your haste to triumphantly rush from the stall and high-five your co-workers to celebrate this most momentous achievement, do not forget to wash your hands! Stool contains between 150 and 500 different types of bacteria in a concentration of 1,012 bacteria per gram! Even in the rare instance of the Wipeless Poo, there are sure to be a couple billion microscopic bacteria on the toilet paper (not to mention the mélange of microbes on the flush handle, the bathroom door, etc.). Some stooling conservatives have advocated for a change in nomenclature, arguing that "Clean Sweep" lends a false sense of hygienic security.

Poo-phoria

Synonyms: *Holy Crap, Mood Enhancer, The Tingler*

This poo can turn an atheist into a believer and is distinguished by the sense of euphoria and ecstasy that you feel throughout your body when this type of feces departs your system. The exhilaration from this defecation, large in volume but varying in form, is often accompanied by goose bumps and even a little lightheadedness as the discharge of the toxins is completed.

You feel energized, as if you just woke up from a great nap. To some it may feel like a religious experience, to others like an orgasm, and to a lucky handful it may feel like both. This is the type of poo that makes us all look forward to spending time on the toilet.

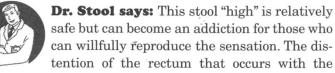 **Dr. Stool says:** This stool "high" is relatively safe but can become an addiction for those who can willfully reproduce the sensation. The distention of the rectum that occurs with the passing of a large mass of stool causes the vagus nerve to fire. The net effect of this is a drop in your heart rate and blood pressure, which in turn decreases blood flow to the brain. When mild, the lightheadedness can lend a sense of sublime relaxation (the "high"). A more significant drop in brain perfusion can cause "defecation syncope," a dangerous syndrome that results in a transient loss of consciousness (the O.D., or Over-Doodie).

CHAPTER 2:

A Butt-Owner's Manual

All most of us can do is find it with both hands, but it's a complex and amazing piece of machinery. Here are a few facts about caring for your hard-working heinie.

The Sit vs. The Squat

For many of us, the act of squatting to "do the deed" conjures up images of less-fortunate children living in societies that lack modern plumbing. For others, stooping to poo has a more nostalgic significance, serving as a reminder of those quick dashes into the bushes for a country dump while camping.

The act of squatting, while seemingly rudimentary, can be fraught with great danger. As the novice squatter begins his descent, he becomes increasingly preoccupied with maintaining his balance while simultaneously ensuring that the stool's trajectory will avoid his pants, socks, and shoes. Inexperienced squatters who are accustomed to the 90-degree angle formed while sitting on the toilet may attempt to re-create this formation during squatting. Unfortunately, this leaves a great distance between poo deployment and landing and dramatically increases the risk of inadvertent lower-extremity soiling. An analogy can be drawn from the world of skydiving, where it is commonly taught that the higher you are when deploying a parachute, the less predictable your landing site
will be.

Bottom line? When forced to squat, low is the way to go.

 Dr. Stool says: Squatting is actually a more effective way of expelling stool than sitting on a toilet seat. A quick review of the physiology of defecation will illustrate this. The rectum is the storage site for stool prior to evacuation. It normally has a tortuous and windy course (picture San Francisco's Lombard Street) that must undergo straightening in order for stool to make its way through the anal canal prior to expulsion from the body. The act of squatting changes the orientation of a group of muscles called the levator ani, which actually serves to stretch open the accordion-shaped rectum. This creates a "straight shot" for the poo to effortlessly make its way out of the body. Just as some naturalists have touted the benefits of breast-feeding and natural childbirth, there is a burgeoning movement that believes squatting can improve digestive health. Not to mention the fact that squatting, also known as the bombardier method, will give you quads of steel.

●●

Nuggets: *In 1987, Dr. Berko Sikirov published a study showing that hemorrhoid sufferers had regression of their hemorrhoids when they went from using sitting toilets to squatting.*

I Have Hemorrhoids, You Have Hemorrhoids

Though the word "hemorrhoid" usually refers to a painful irritation of the veins around the anus, the word actually just means the veins themselves. In other words, everyone has hemorrhoids, and they're perfectly harmless. The proper name of the painful condition is "hemorrhoid disease."

Hanging Chad

Synonyms: *Grundle Weeds, Cling-Ons, Butt Bark, The Lone Ranger, Turtle Head, Crap Crumbs*

It is critical when wiping up after a poo to make sure the job is complete. Hanging Chads are the stubborn pieces of turd that cling to the anal hairs and often refuse to let go. Larger than the traditional dingleberry, an unnoticed Hanging Chad can provide an unwelcome surprise when you remove your underwear or take a shower. While some find these surprises humorous, they can be disastrous if your lover is the one who discovers these leftovers. When such a calamity occurs, not even the U.S. Supreme Court can step in to save you. In order to dislodge the Lone Ranger, one may need to rock back and forth or utilize a bouncing motion at the end of the bowel movement. The last recourse, should these poo calisthenics fail, is to find a healthy piece of toilet paper (preferably double-ply) and utilize the unappealing, yet surprisingly effective, "pinch" maneuver. Tempted to obliterate the hanging chad by wiping more vigorously? Remember the old adage: "Fear the smear."

●●●●●●●●●●●●●●●●●●●●●●●●●●●●●●●●●●●●●●

Nuggets: *Speaking of cling-ons, the Klingon words for poo are "baktag" and "HoH."*

Smoke Signal

Synonyms: *The Sentinel, Opening Act, The Warning Shot, A-poo-tizer*

Seemingly just another fart upon deployment, the Smoke Signal is noteworthy because of what it portends. Its release typically occurs thirty minutes to one hour after a meal and can vary in intensity from a spurt to an eruption. Its aroma, however, is consistently abhorrent and can initially cause the responsible party to fear that he has expelled more than just gas. It may even prompt a quick pat to the backside to ensure that no solid matter has inadvertently escaped. A dash to the loo quickly follows the conclusion of the Smoke Signal, leaving little doubt in the minds of onlookers as to the perpetrator of this especially pungent "Opening Act."

 Dr. Stool says: The Smoke Signal is best thought of as foreshadowing, an evolutionary advance that alerts the owner (and, sometimes, others nearby) of the need to take care of more serious business. The release of a Smoke Signal broadcasts to those around you your need to poo. The abhorrent aroma is the result of its prolonged cohabitation with fecal material. Rich in odoriferous sulfides, the Smoke Signal smells identical to a freshly laid log of good ol' number two. Its aroma is far from pleasant, and while denial is the usual reaction, ignoring its call for action can be even more catastrophic.

CHAPTER 3:

Everything You Ever Wanted to Know about Toilets

The toilet is where you take a load off, in every sense of the word. It's the most important thing in the house, and we couldn't be more grateful for it. Maybe that's why we end up kneeling in front of it so often.

The History of Toilets

Toilets date back to at least 3000 BCE. Evidence of primitive toilets has been found in the remains of neolithic Scottish settlements. By 2500 BCE, the Harappan civilization had indoor water-flushing toilets in what is now Pakistan and India. Toilets existed in Minoan Crete, ancient Persia, ancient China, ancient Egypt, and ancient Rome. The toilets found in Roman bathhouses provide some of the earliest examples of sitting toilets, which didn't come into general use until the mid-1800s.

Before modern plumbing, however, the restroom was not where you went for some "me time." Toilets could be scary and disgusting places, and visitors would turn to the supernatural for protection. Archaeologists excavating ancient Roman lavatories often find protective charms written on the walls.

●●

Nuggets: *The little bumpers under your toilet seat actually have a name. They're called buculets.*

Testing, Testing, One Two

Ever wondered what toilet manufacturers use to test their products? Instead of real poop, they flush them with fermented bean curd. Brown and soft, it looks and behaves like the real thing, clogging just as poop would.

DOO YOU KNOW?

HOW MANY POOPS DOES IT TAKE TO FILL A PORT-A-POTTY?

An average portable (Port-a-Potty) toilet is able to hold enough sewage for 10 people during the course of a 40-hour work week before it reaches unsanitary conditions.

Can Sitting on the Toilet for Too Long Cause Hemorrhoids?

Yes. Medical studies have found that if you spend an excessive amount of time on the toilet, you may be at risk for developing hemorrhoids (regardless of how smoothly you complete the deed). In fact, one study even found that hemorrhoids became smaller after patients stopped reading on the toilet! It seems that spending too much time on the toilet can be bad for your health.

TOILET FACTS

- The toilet uses more water than any other home appliance.

- The life expectancy for a toilet is 50 years—about the same as for a gorilla.

- In a public bathroom, the first stall is generally the least used, and thus, the cleanest.

- Most toilets flush in E flat.

Gift Poo

Synonyms: *Pollyanna Poo, Phantom Poo, Surprise Party, Shock and Awe, Flushless Folly, River Pickle*

Gift Poos are turds that people leave behind in toilets without flushing. They come in all shapes and sizes and are most prevalent in public restrooms and fraternity houses. These gifts are sometimes left as trophies to anonymously show off accomplishments, or as pranks, or occasionally as absent-minded accidents (often resulting from multitaskers trying to talk on their cell phone and wanting to avoid the flush that gives them away). The origins of these poos may remain a mystery, but they can linger anywhere from minutes to days, depending on when the gift is first seen or smelled.

Dr. Stool says: Stool can remain for days, even weeks, submerged in the friendly confines of a toilet bowl. Whereas some turds retain their cohesion and luster for extended periods of time, the usual Gift Poo will gradually fragment and liquefy.

The Streak

Synonyms: *Skidmark, Hershey Highway, Racing Stripe, Lining the Pavement*

A phrase more commonly reserved for Joe DiMaggio's seemingly unbreakable fifty-six-game hitting streak in 1941, The Streak also has a well-established place in the world of poo. The Streak is a relic of a prior poo usually appearing as a thin brown stain down the center of the toilet bowl. Some streaks maintain their legacy and remain visible for multiple flushes after their original introduction to the toilet.

The appearance of a racing stripe at the bottom of the bowl is a most unwelcome sight for all would-be poopers, especially guests at a friend's dinner party. To ensure you don't leave a poo trail that leads to you, always give the post-flush glance to make sure you don't need an encore flush.

 Dr. Stool says: While this is no streak to be proud of, it is rarely a cause for concern. The appearance of The Streak is highly unpredictable, and there is no evidence to suggest that specific foods are responsible for lending this stool its sticky nature. One potentially worrisome scenario associated with thick, sticky stools is upper gastrointestinal bleeding. In this case, blood originating from "high up" in the GI tract (i.e., the stomach) is transformed during its passage through the intestines into a thick, tarry stool that is usually jet black in color and extremely foul smelling. A dramatic change in the color of stool (to either black or red) can often be the first indication of serious gastrointestinal bleeding.

Itchy Poo

Synonyms: *Scratch and (Don't) Sniff Stool, Prickly Poo*

Everyone, at some point, has had an uncontrollable, irrepressible urge to scratch the anal region after having a bowel movement. The urge to itch commonly comes as we lie in bed at night, sometimes during sleep itself. For most, this sensation will quickly pass within a day or two. For others, the desire to grab the nearest emery board and go at it persists for weeks and months. When occurring at night, this condition can interfere with restful sleep and leave nighttime scratchers tired and sluggish the next morning. When occurring during the day, the anal scratching will no doubt offend classmates, colleagues, and passersby. Public humiliation aside, anal itching can cause underwear to become stained and fingers to smell. If a quick change in your diet and laundry detergent doesn't offer relief, it's time to do yourself (and others) a favor and walk yourself, hands firmly in your front pockets, to the nearest doctor.

Dr. Stool says: The most common causes of perianal itching are medications (especially antibiotics), chemical irritants (laundry detergents, lotions), overly aggressive wiping, hemorrhoids, and a pinworm called *Enterobius vermicularis*. The itching associated with this pinworm infection is especially fierce at night when the female parasites emerge from the anal canal to lay ten thousand or so eggs on the perianal skin. These eggs cause severe irritation to the area, leaving the unsuspecting host no choice but to scratch away, thereby transferring eggs to the hands where they can be spread to others.

CHAPTER 4:

The Wonderful Whiz

Think of pee as your body's way of giving its insides a bath. It's brilliant! No wonder we call it Number One.

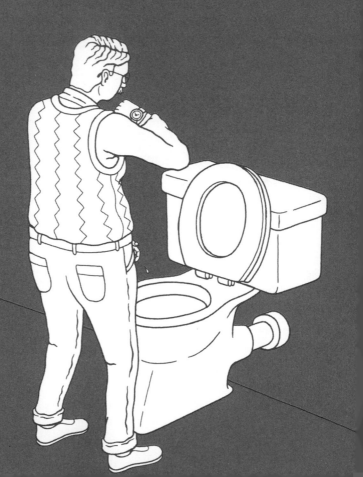

What's in Your Pee?

As you'd expect, pee is mostly water, usually 91 to 96 percent. The remainder is made up of salts, proteins, hormones, and waste from whatever your body has metabolized, be that alcohol or asparagus.

How to Get Your Toga Its Whitest—and Your Teeth, Too!

In ancient Rome, clothes were laundered in human urine, collected from public latrines. Urine contains ammonia, a natural whitener. The clothes were then rinsed in water to remove the urine smell. The process was quite effective for cleaning the clothes but not much fun for the workers at the laundry, or fullonica, who had to spend the day standing in vats of urine stomping on pee-soaked garments.

Ancient Romans used pee not only on their laundry but on their teeth. Urine was sometimes used as a mouthwash to clean and whiten. For oral use, Romans preferred imported urine, and gargled with pee bottled and brought in from Portugal. Mouthwashes continued to use urine into the eighteenth century.

Fountain of Youth

In 1978 Indian prime minister Morarji Desai made a startling admission to Dan Rather on *60 Minutes*: he drank his own urine. Desai was an avid practitioner of "urine therapy," calling it an effective treatment for Indians who could not afford other medical care. Desai was widely ridiculed for the statement, both in India and abroad. But he did live to nearly 100, so maybe he was on to something.

●●●

Nuggets: *If consumed on an empty stomach, urine can act as a laxative.*

Dr. Stool Says: Should You Drink Your Own Urine to Survive Extreme Dehydration? Sometimes. There is ample anecdotal evidence of people surviving in life rafts or trapped in collapsed buildings by drinking their own urine. To determine the safety of drinking urine, remember two things: urine is 95 percent water (at least for the first few days of dehydration) and it is sterile. Our bodies need water to survive (we can go weeks without food), so urine is a surprisingly logical choice when no other hydration options are available. The jury is out on whether this practice is safe for more than a few days, as urinary concentration of harmful substances (e.g., urea) increases over time.

Pebble Pee

Synonyms: *Tortured Tinkle, Agony Pee*

The passage of pebble pee is as excruciatingly painful as childbirth—without an epidural. More often occurring in men, Pebble Pee often begins as just another trip to the bathroom. The onset of sharp groin pain (similar to labor pains) is the first sign that things are about to get interesting. What usually follows is a several-hour-long melee filled with writhing, perspiration, and profanity. Only with the passage of the offending kidney stone does the agony end. Unlike the joyous results of pregnancy labor, the only fruit of your agony is a two-millimeter pebble resting tauntingly in the toilet bowl.

 Dr. Stool says: The passage of a kidney stone is among the most painful of all bodily processes. Formation of stones occurs when the urine contains high concentrations of certain substances (most commonly calcium). These substances begin to precipitate and form crystals, which then coalesce to form stones. Stones can become lodged anywhere along the urinary tract, causing excruciating pain as urine flow becomes blocked and muscles contract more forcibly in an attempt to overcome the obstruction. Depending on the site of obstruction, pain may be felt anywhere from the back to the groin to the urethra. Increasing urine flow by drinking large quantities of water and using pain medications are the usual treatments while awaiting spontaneous stone passage. In some cases, surgical intervention is needed to remove large stones too big to pass through the urethra.

Pee-phoria

Synonyms: *Pleasure Pee, Piss Bliss, Ecstasy Pee*

Sometimes we are forced to hold in pee, typically on a long car ride or in an important meeting. What begins as a small urge rapidly becomes an unbearable discomfort, with each bump in the road serving as a reminder of your overloaded bladder. When finally able to stand and head for the toilet, some will engage in what is commonly referred to as the "Pee Pee Dance" or "Tee Tee Tango," a series of random body gyrations that may include swaying back and forth, bending over and standing up, even hopping up and down (although this seems counterproductive) before the eventual release of pee. As urine begins to flow, this agony gives way to utter elation. Some have described a cool, tingling sensation that radiates throughout the body as the urine is discharged.

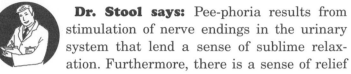 **Dr. Stool says:** Pee-phoria results from stimulation of nerve endings in the urinary system that lend a sense of sublime relaxation. Furthermore, there is a sense of relief as bladder pressure diminishes, leaving you feeling light on your feet. In most cases, there is no danger in holding in urine for a short while. Pee-phoria's intensity is directly proportional to the degree of agony experienced while holding it in.

Mellow Yellow

Synonyms: *Gold Rush, Sunny P*

No color may have more positive associations than yellow. Yellow sunflowers are widely considered to be some of the most cheerful flowers on earth. The yellow jersey represents victory in the Tour de France. Yellow smiley faces are the symbol of happiness. And, yes, urine, too, has always been associated with the color yellow. Urinophiles love to wax poetic about pee's sparkling, golden appearance—seventeenth-century alchemists were convinced that the yellow color was due to the presence of gold in urine. While a yellow-tinged hue to urine is normal, darker, richer yellows are less desirable. In the world of pee, bolder is not better: mellow yellow is the way to go.

 Dr. Stool says: Urine's yellow color comes from urobilinogen (not gold), which is a break-down product of bilirubin (itself a breakdown product of hemoglobin). Bili-rubin enters the kidneys where it is converted to yellow-hued urobilins and excreted in the urine. Assuming normal levels of urobilins, the color intensity of urine is determined simply by the amount of water passing through the kidneys.

Dehydration causes the kidneys to retain water for the body's use, so less water ends up being excreted into the urinary stream, thereby concentrating urobilins and rendering urine a darker color. Simple tests can measure the concentration of urine by assessing something called the specific gravity. In general, the higher the specific gravity, the darker the urine will appear. In fact, urine color can be used as a rough gauge of the body's hydration status during episodes of food poisoning or after a long-distance run. Dark yellow urine indicates concentrated urine, signaling the need to consume more water.

DOO YOU KNOW?

UP AND DOWN

The pH level of urine fluctuates, with the first pee of the morning being the most acidic. Urine grows more alkaline through the day, as food is processed and electrolytes are released.

Vitamin Water

Synonyms: *B Pee, Vitamin P, Liquid Highlighter*

While the normal palette of pee can range from completely clear to bright yellow, there are times when pee comes out a fluorescent yellow, or even a bright orange. The seemingly unnatural vibrancy makes this the most dramatic of pee's color variations—its brightness can leave you searching for the nearest pair of sunglasses. Some people are tempted to turn off the lights and see whether the toilet water glows in the dark. When you experience Vitamin Water pee, you may start to wonder if your diet has included antifreeze or highlighter ink.

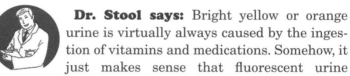 **Dr. Stool says:** Bright yellow or orange urine is virtually always caused by the ingestion of vitamins and medications. Somehow, it just makes sense that fluorescent urine wouldn't occur as a result of any naturally occurring bodily process. Vitamin B supplementation has become popular due to its potential to boost the immune system and to promote healthier skin. When taken in large doses, the body disposes of the excess vitamins in the urine, lending it a bright yellow/orange color. Other medications including senna (used to treat constipation) and pyridium (interestingly, used in treatment of urinary tract infections) are well-known culprits of orange-hued urine. The change in urine color is not harmful in any way, but may be a sign that you are taking more vitamins than your body needs.

Gravy Train

Synonyms: *Brown Bonanza, Chocolate Drizzle, The Coca-Cola*

When seeing brown urine for the first time, you may attribute the toilet water discoloration to an ineffective flush following a bowel movement. As you begin to curse your roommate for not monitoring the toilet bowl for poo remnants, you glance down and notice a distinct cola-like quality to the urinary stream as it emerges from your body. The Gravy Train makes one wonder whether it's possible to pee out poop.

 Dr. Stool says: Occasionally, it may seem as if the body has its signals crossed, turning urine brown and stool white. Brown pee is caused by liver dysfunction or a blockage in the bile ducts that causes spilling of bile into the bloodstream and then into the urine. Bile lends urine a dark greenish-brown color, classically described as being "Coca-Cola-colored" or "tea-colored" in appearance. Blockage of the bile duct caused by a gallstone also prevents bile from entering the intestinal tract, thereby depriving stool of its usual chocolate color. Because bile duct blockage can also result from pancreatic tumors, the combination of dark urine and pale-colored stools should always prompt a visit to the doctor.

DOO YOU KNOW?

PORPHYRIA

Porphyria, believed by some to explain the erratic behavior of King George III, is caused by defective enzymes involved in making heme (a component of hemoglobin). Diagnosis of this rare disease can be made by observing a fresh urine specimen turn from yellow to red when exposed to sunlight.

The Drip

Synonyms: *Prostate Pee, Faulty Fountain, Sputtering Hose*

Men: To the list of things that slow down as you age, add the speed of your urine stream. As young boys, we engaged in distance-peeing contests, shamelessly flaunting our ability to forcefully propel our urine long distances. Almost universally, this once-robust jet stream will eventually give way to frustrating drips and drabs more akin to a leaky faucet than a fire hose. Despite forceful straining and the occasional milking maneuver, urinary flow can deteriorate to the point of complete obstruction. The inability to satisfactorily empty the bladder leads to frequent, frustrating trips to the bathroom and multiple nighttime awakenings to visit the loo.

 Dr. Stool says: Normal urinary flow is a function of the strength of bladder contraction versus the degree of flow resistance in the urethra. Normal urinary flow rate is dependent on age and gender but typically ranges from ten to twenty milliliters per second. Enlargement of the prostate gland, a normally walnut-sized structure which encircles the urethra, is a nearly universal occurrence as men age. The enlarged gland can compress the urethra and cause an increased resistance to urinary flow, thereby decreasing urinary stream velocity. Mild cases of prostatic hypertrophy can be treated with medicines that shrink the gland. More severe cases can cause complete obstruction to urinary flow, necessitating the surgical placement of a urinary catheter.

CHAPTER 5:

It's a Gas, Gas, Gas

Farts are an ephemeral phenomenon, here for but a moment. Life moves fast; don't forget to stop and smell them.

The Major Components of Human Farts Are:

Unicorn farts are made of glitter and wishes. Human farts, however, are made of more organic materials. The typical fart contains:

- Nitrogen: 20 percent to 90 percent
- Carbon Dioxide: 10 percent to 30 percent
- Hydrogen: 0 to 50 percent
- Oxygen: 0 to 10 percent
- Methane: 0 to 10 percent

Is Holding in a Fart Bad for You?

Ever since humans began living and working indoors, our ability to fart with impunity has been compromised. No longer can we let 'er rip without first ensuring our isolation. As a result, despite the discomfort, we sometimes elect to hold it in. Flatus retention is frequently associated with some transient abdominal distress, but it is only natural to wonder whether there is any serious downside to ignoring the body's urge to vent.

Some flatus experts have linked increasing rates of diverticulosis (outpouchings in the colon wall) to flatal retention. Holding air in the colon, they reason, increases intracolonic pressure, causing herniation and the formation of diverticula over time (like overinflating an old tire). No studies have proven this association but the rates of diverticulosis have been shown to be higher in industrialized countries where, presumably, office workers have greater pressure to suppress gaseous emissions.

Fighting Global Warming, One Fart at a Time

There may be one benefit from societal pressure to suppress farts when surrounded by others: less global warming! Methane gas makes up a small, but sizeable, proportion of gases emitted with each fart. Its ability to trap heat in the atmosphere makes it a significant contributor to global warming. The importance of methane destruction in the fight against global warming is best demonstrated by the ongoing research dedicated to reducing the amount of methane found in cow flatulence by modifying their diets.

Foods that Cause Gas

Maybe you're trying to avoid gas. Maybe you're trying to achieve it. Here are the foods that'll do it for you:

- Beans
- Vegetables, such as broccoli, cabbage, Brussels sprouts, onions, artichokes, eggplant, and asparagus
- Fruits, such as pears, apples, apricots, prunes, and peaches
- Whole grains
- Soft drinks, dark beer, instant coffee, and red wine
- Milk
- Sorbitol (used as a sugar substitute in diet foods)
- Molasses

Gender Gap

An Australian study found that men fart an average of ten times a day while women let loose a mere eight times a day. The study also concluded, however, that women's farts smelled worse, mainly because female farts had a higher sulfur concentration. How were the fart smells analyzed? First, subjects farted into an aluminum bag via a rectal tube. The gas was then removed with a syringe and expelled into the nostrils of the eagerly awaiting judges. All in the name of science!

DOO YOU KNOW?

FART CHEMOTHERAPY

Can smelling farts cure cancer? Almost certainly not—but maybe! A study done at the University of Exeter showed that sniffing hydrogen sulfide—the smelly component of farts—can prevent mitochondrial damage, thereby reducing your risk for cancer as well as arthritis, stroke, and dementia.

Morning Thunder

Synonyms: *Wake-up Call, Revelry, The Rooster*

This rumbling and resonant riptide of flatus heralds the beginning of a new day. Air passage typically occurs within thirty minutes of awakening, and if your partner or roommate happens to be asleep, the conch-like bellowing sound is sure to wake them. Admittedly not the most soothing way to be awakened, having a roommate with potent and predictable Morning Thunder does have the advantage of precluding the need for an alarm clock. In addition to its typical deep, reverberating sound, these early AM emissions are long in duration, easily lasting three to four seconds. Thankfully, this flatus is all bark and no bite; Morning Thunder's odor is classically benign.

 Dr. Stool says: Morning Thunder is the result of an increase in the colon's muscular contractions which occur upon awakening. We all pass small amounts of gas while we sleep, but for most of the night, our colons are resting quietly. Upon awakening, our colonic activity increases dramatically, causing rapid propulsion of fecal matter and gas along the length of the colon to the rectal area. These tidal-wave-like movements, called HAPCs, or high-amplitude propagating contractions, account for the need to break wind before we breakfast. Some individuals experience mild abdominal discomfort as gas is rapidly propelled down the colon, but quickly achieve relief with this early morning decompression.

Silent But Deadly

Synonyms: *Stealth Bomber, Silent But Violent, Elephant in the Room*

We've all been there. Thirty minutes into the post-lunch staff meeting you begin to feel the rumblings of impending flatulence. Confident in your ability to release this air bolus in a covert fashion, you decide to liberate a small, seemingly harmless waft of air right there in the boardroom. You feel a sense of relief after the successful deployment and begin to grow proud of your accomplishment, marveling at your Zen-like ability to stealthily decompress your colon. Several seconds later, this braggadocio is replaced by sheer horror as you realize this ephemeral emission has taken on the aroma of a toxic waste dump. You may look around to see if there is a dog in the vicinity that you can blame. Ultimately, your fear is realized when colleagues begin shifting in their seats, lips firmly sealed, their faces scrunched in revulsion. You quickly feign the same sort of disgust, even mumbling to your neighbor, "Aw, man . . . who cut the cheese?" lest you be identified as the source of this most atrocious atmospheric contribution.

Dr. Stool says: The SBD is proof that while one can often estimate the volume of an impending fart, its odor can be infuriatingly unpredictable. In fact, a fart's toxicity has surprisingly little to do with its volume and everything to do with its concentration of hydrogen sulfide. Hydrogen sulfide, not the notorious methane, is the ingredient which lends flatus its distinctive eggy aroma. This gas is formed when bacteria ferment raffinose, a particularly potent carbohydrate found in brussels sprouts, cabbage, and other flatus-inducing plant-based products.

Mile High Club

Synonyms: *High-Flying Farts, Airlift, Altitoots*

Sir Edmund Hillary and his Sherpa, Tenzing Norgay, are considered great men for their ability to overcome challenging conditions during their ascent to the top of the world's tallest mountain in 1953. The rugged terrain, ice storms, and bone-chilling temperatures must have paled in comparison to a lesser-known challenge of mountain climbing: flatulence. These heroic men almost surely suffered from excruciating abdominal pain resulting from the enormous quantity of gas which formed in their colons as they ascended Mount Everest. When it comes to flatulence, climbing to 29,000 feet is the equivalent of consuming bowl after bowl of black bean chili. A rarely discussed hazard of mountain climbing, the buildup (and release) of this expansive flatus can cripple even the most seasoned of climbers, making Hillary and Norgay's accomplishment even more remarkable.

CHAPTER 6:

Diarrhea, Diarrhea

When you're reading about poop, and your unders fill with goop . . . Diarrhea! Diarrhea!

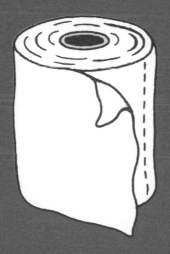

Dys and Dat

What's the difference between dysentery and diarrhea? Although the terms are often used interchangeably, diarrhea and dysentery are different conditions. Diarrhea affects the small bowel and is characterized by watery stool. Dysentery affects the colon and contains mucus and, typically, blood. It is often accompanied by fever and cramping of the lower abdomen. More serious than diarrhea, dysentery requires medical treatment.

Orange You Glad You Don't Have It

Though it sounds like a pretty first name, keriorrhea is actually a type of diarrhea characterized by greasy orange stools, which doesn't sound so pretty at all. Keriorrhea is caused by eating certain types of fish, like escolar and oil fish, that contain indigestible wax esters.

DOO YOU KNOW?

FACTITIOUS DIARRHEA

It sounds made-up, but "factitious diarrhea" is a real medical term that describes deliberately self-inflicted diarrhea, usually caused by laxative abuse. It is seen more often in women than in men, and is more common among the well-off.

Hometown Pride

The only town in the United States that's given its name to a type of diarrhea is Brainerd, Minnesota. Brainerd diarrhea is a type of acute and severe diarrhea that doesn't respond to antibiotics and can last for months. It can occur individually or in outbreaks, and while the cause is unknown, it's associated with the consumption of raw milk and untreated water. Although it's been observed in ten different areas, it was named after the town where it was first seen, in 1983.

Lactose Intolerance: An Evolutionary Tale

A worldwide study of populations with and without lactose intolerance sheds some interesting light on the condition's history. In ancient China, nomadic Mongolians consumed large quantities of horse milk because they never settled down long enough to turn the milk into cheese (which has very low levels of lactose). Inland residents of ancient China, by contrast, consumed very little milk as they preferred consumption of mature cheeses. These differing practices thousands of years ago have led to widely divergent rates of lactose intolerance in modern China—very low in Mongolian populations and very high in the Chinese.

Soft Serve

Synonyms: *Jabba the Poo, Play Doo, Cow Pattie, Septic Seepage*

More dense than diarrhea but softer than a normal poo, this solid yet amorphous turd comes out in one smooth, steady flowing motion. Its easy exit may make you feel like the stool will take a liquid form, but you are pleasantly surprised to see its more cohesive consistency when you are done. While still far from a cylindrical shape, these poos look a lot like cow patties. If

you deposited this poo directly into a bowl, it would be easy to mistake for a dish of soft-serve ice cream from Dairy Queen on a hot summer day (although the odor would help make the distinction quite clear). This form of poo sometimes comes as a precursor to, or a last stage of, some sort of intestinal disturbance.

Dr. Stool says: Stool's liquid consistency can be intermittent, often varying from one bowel movement to the next. When short-lived, the development of mildly loose stools associated with abdominal cramping is usually due to ingestion of poorly absorbed foodstuffs. While substances such as fructose (in juices and sodas) and sorbitol (in sugar-free gum) are becoming increasingly recognized, the most common cause of the Soft Serve is lactose intolerance.

Lactose intolerance—the inability to digest the main sugar found in milk—is a remarkably common condition worldwide, especially within certain ethnic groups. Asian and African American populations have rates of lactose intolerance that approach 80 percent. Lactase, the enzyme responsible for breaking down lactose, is found exclusively in the small intestine and is affected by a variety of conditions. While most lactose-intolerant individuals are genetically deficient in lactase, conditions such as inflammatory bowel disease, which injure the small intestine, can also reduce levels of the lactase enzyme. Lactose intolerant and craving that extra-thick milk shake? Try taking supplemental lactase enzyme pills to help process the load. Another way to enjoy your morning cereal if you are lactose-intolerant is to substitute soy milk, which has the added benefit of adding protein to your diet.

If you know your Soft Serve is not being caused by lactose intolerance, you should ask your doctor if you have celiac sprue, a disease characterized by an intolerance to gluten-containing foods. Gluten is a component of grains such as wheat, rye, and barley, and ingestion by susceptible individuals causes an immune reaction that damages the small intestine. Crampy abdominal pain, watery diarrhea, and the eventual development of an iron deficiency (anemia) are common adult manifestations of this condition. Diagnosis can be made by blood tests and upper endoscopy. Treatment requires the elimination of all gluten from one's diet.

More often than not, this form of poo does not require specific treatment. An increase in dietary fiber will help to bulk up the stool and lend it a more traditional consistency.

Number Three

Synonyms: *Butt Piss, Liquid Poo, The Runs, Oil Spill, Hershey Squirts, Montezuma's Revenge, Chocolate Thunder, Diarrhea, Operation Marination, Operation Evacuation, Releasing the Hounds, The Nile, Poo Stew, Chocolate Slurpee, Gravy Poo, Birds Flying South for the Winter, Rectum Rapids, It's Raining Poo, Deuce Juice, Turd Tea*

Although you know that you need to sit down for this rear deposit, Number Threes come in a liquid form and have little to no texture. When passing one, you feel as if you are urinating from the wrong side. A Number Three is often a violent discharge, sometimes with very little warning, and may often be accompanied by tremendous gaseous emissions. As you feel its sudden onset, your sense of relief that you made it to the toilet in time is quickly replaced by the ill feeling associated with the release of a Number Three. The explosiveness is so severe that it often results in brown splatter hitting the underside of the toilet seat. At times, the splatter is so great that you have to wipe remnants off your butt cheeks when you are finished. Number Threes are not pleasant.

Dr. Stool says: The Number Three has two main causes: GI tract infections and maldigestion. Inadvertent consumption of bacteria, viruses, or toxins from undercooked meat or week-old potato salad causes the small bowel to secrete large volumes of fluid into the GI tract. This deluge of fluid, coupled with brisk intestinal transit (picture the torrent of Class 5 rapids), results in the delivery of large amounts of liquid to the rectum. Cholera infection is the most severe example of this physiology. The diarrhea produced by this disease is classically referred to as having the consistency of "rice water" and leads rapidly to life-threatening dehydration.

Impaired digestion is the other possible cause of the Number Three. One should consider this diagnosis when the explosiveness of the bowel movement is particularly violent. In lactose-intolerant individuals, ingestion of dairy products results in the production of copious amounts of gas and liquid stool. If severe enough, the expulsion of these "contents under pressure" can cause your significant other to run for cover.

D.A.D.S.

Synonyms: *Revenge of the Poo, Morning After, Poo of Shame, Bud Mud*

A D.A.D.S. is a day-after-drinking stool. After a long night of partying, you may awake the next day with a hangover and an unsettled stomach. Whether your personal hangover cure is a greasy meal, a milk shake, a Bloody Mary, or a family elixir, your body needs to purge itself of the toxins and recover from the indulgences of the previous evening. A D.A.D.S. often comes in a semisolid state, and sometimes is accompanied by stomach discomfort. The most notable trait of a D.A.D.S. is the tread mark left on the toilet bowl after you flush, as well as the distinctive bar-floor smell. The more you drank the night before, the more D.A.D.S. you need to eliminate to start feeling better. Usually your second D.A.D.S. of the day signifies that your recovery is well under way.

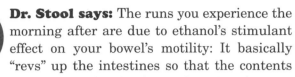 **Dr. Stool says:** The runs you experience the morning after are due to ethanol's stimulant effect on your bowel's motility: It basically "revs" up the intestines so that the contents move through more quickly. This leaves less time for your colon to absorb water, and results in a profuse, watery stool. Occasionally, the large carbohydrate load in alcoholic beverages can overwhelm your digestive enzymes and indirectly cause diarrhea.

Poo-nami

Synonym: *Baby Blowout*

Diaper technology has improved over the decades, but there has yet to come an innovation that can suppress the Poo-nami. While most baby poos can be contained by a diaper, there are instances when nothing can stop the poo from exploding beyond the confines of a diaper, up the baby's back, and down his or her legs. Dealing with Poo-namis is just one of the many challenges of parenthood.

Dr. Stool says: Ill-fitting diapers are the most common cause of the dreaded Poo-nami. New parents who are inexperienced in proper diaper application techniques often suffer their share of Poo-namis during the first few months of parenthood. Poo-namis also tend to occur during bouts of diarrhea, most commonly due to rotavirus infection.

It is quite normal for infants to have explosive, liquid bowel movements in the first few months of life. This occurs because stool is frequently expelled simultaneously with gas. When stools become very frequent and voluminous, diarrheal illnesses such as rotavirus infection may be to blame.

CHAPTER 7:
In a Bind

Constipation is no fun. When you're backed up, all you want is to get back in the saddle. Hope you find a happy trail soon, cowboy.

Go with the Flow

Poo's delivery should be effortless—more akin to yoga than power lifting. More important than generation of high levels of abdominal pressure is the relaxation of the pelvic muscles and anal sphincter. Uncoordinated contractions of the muscles in the rectal area result in a condition known as pelvic floor dyssynergia, in which anal muscles contract instead of relaxing, thereby obstructing the flow of stool. Pushing with more force is the typical response, but this only worsens the constipation. Treatment is aimed at teaching patients to relax through a process called biofeedback.

While it may come as a surprise to some, increasing your fiber intake can sometimes make constipation worse. Sure, it bulks up the stool, but unless you consume enough water, your stool may actually become harder and more difficult to pass. So when fighting for a more peaceful poo, be sure to drink six to eight glasses of water a day in addition to eating high-fiber foods. If the combination of more fiber and a lot of water does not help cure constipation, the next step would be to visit your local GI specialist to discuss other tests and the possible use of laxatives (medicines that speed up the movement of the intestines).

Hot Buttered Enema

The diarist Samuel Pepys used to treat his constipation with an enema whose ingredients included butter and sugar. He claimed to be quite pleased with the results.

The Worst Case of Constipation

The world's largest preserved colon once contained over 40 pounds of stool and was stretched to many times its original size. On display at the Mutter Museum in Philadelphia, the colon's owner suffered from Hirschprung's disease, a thankfully rare disorder which results in the inability to empty one's bowels.

●●

Nuggets: *Constipation comes from the Latin word* constipare, *meaning "to crowd together."*

Fecaloma

A fecaloma ("fecal tumor") is a hard mass of stool that forms in the colon as a result of conditions like Hirschsprung's disease or prolonged chronic constipation. Also known as fecoliths or coproliths, fecalomas can grow quite large, in which case they have to be removed surgically.

Slowing Down

Why does Grandpa spend so much time in the bathroom? Because constipation becomes more common as we age. It's also more common in women than men, so expect Grandma to take even longer.

Get Things Moving

If you're prone to constipation, you may want to steer clear of these constipating culprits:
- Milk, cheese, and yogurt
- Bananas
- Cooked carrots
- White rice
- Processed and high-fat foods like French fries

Log Jam

Synonyms: *Can't Get the Train Out of the Tunnel, Colon Congestion, Corked, F.O.S., Stuck Up, Crying Wolf, False Alarm*

Despite stomach pains, rancid gas, and feeling a turd on deck, no matter how hard you push, nothing comes out. After ten to fifteen minutes in the bathroom, your friends, spouse, or roommate may start to worry about you, but you may not be ready to give up yet. However, when you ultimately decide that it was a false call, the emptiness of the toilet bowl is a cruel reminder of your inability to perform.

Dr. Stool says: This is the most feared complication of constipation. A lack of dietary fiber and water can result in a stool bolus so hard and so firm that it is unable to pass through the anal sphincter. Whereas normal stool is able to smoothly and effortlessly exit the rectum, a desiccated boulder-like bolus often cannot escape its confinement without assistance. Treatment entails the administration of enemas and, in severe cases, manual disimpaction. Want to avoid this unpleasant, intrusive manipulation? Dr. Stool recommends that your diet contain an ample amount of fiber and water.

Pebble Poo

Synonyms: *Rabbit Poo, Kibbles 'n' Bits, Splashers, Butt Hail, Blueberries, Buckshot, Meteor Shower*

You may sit down at the toilet with aspirations for a large, enjoyable poo, only to have Pebble Poos leave you unsatisfied and unfulfilled. Despite your vigorous straining and the sensation of poo exiting your rectum, when you stand up to look there are only a handful of pebbles resting mockingly on the toilet bowl floor. Adding insult to injury are the unwelcome splashes that hit your buttocks as the mini poo pellets hit the water.

Dr. Stool says: Pebble Poo reflects a lack of stool cohesion. How does the GI tract produce a well-congealed, singular bolus of soft stool? It uses glue, of course. This "glue" is actually a fatty-acid gel that is formed when ingested fiber is fermented by bacteria residing in the colon. This sticky substance, not unlike the gooey marshmallow mixture in Rice Krispies treats, keeps poo from breaking apart and drying out. This magical gel also lubricates the inside of the colon, allowing the stool bolus to pass friction-free through the GI tract. A lack of dietary fiber results in small, hard, disjointed poos that can give rise to the most un-Zen-like of stooling experiences.

CHAPTER 8:
On a Roll

People have been pooping since butts were invented, but toilet paper only came along recently. What would we do without it?

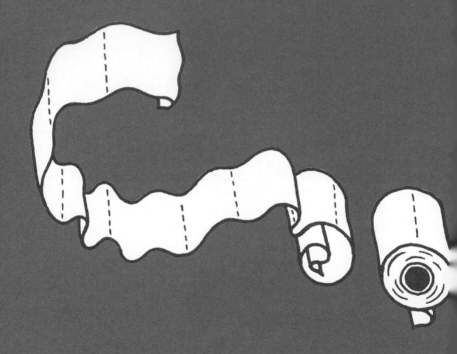

Get That Paper

Toilet paper was first used almost two thousand years ago in China, but it took a lot longer to catch on in the rest of the world. The first packaged toilet paper sold in the United States came on the market in 1857. Sold by the Gayetty Firm, it featured the company name on every sheet. When it was first introduced, toilet paper was considered more of a medical supply than a daily necessity. It was called "therapeutic paper" and contained aloe to treat sores. Sold in pharmacies, it cost 50 cents for a package of 500 sheets, the equivalent of $15 today. Disturbingly, splinter-free toilet paper wouldn't come along for almost eighty years. That was invented in 1935 by Northern Tissue, which came up with a new method called linenizing to remove the small splinters that had previously plagued the product.

The Wipe Stuff

Before the widespread availability of toilet paper many other materials were used to accomplish a proper cleanup. In many parts of the world, the wealthy used animal furs and expensive materials such as lace, while those not as fortunate were forced to use leaves, sand, and even corn cobs. People in coastal areas commonly wiped with clam shells. At sea, sailors wiped with frayed rope—but that's not the grossest part. This is: they all shared one rope!

TOILET PAPER FACTS

- In some countries, toilet paper isn't flushed. If you see a wastebasket full of unsightly wipings next to the toilet, follow suit and dispose of your paper there—it's likely that the plumbing or septic system can't handle paper.

- Toilet paper is most expensive in the United Kingdom, where it costs twice as much as it does in Germany or France, and almost three times as much as it does in the United States.

- Most people use eight to nine sheets of paper per trip to the toilet, adding up to an annual average of 23.6 rolls per capita in the United States.

- An average home will go through a roll of toilet paper every five days.

- Two-thirds of Americans prefer the first sheet to come over the top the roll; one-third prefer it to come under the bottom.

- Seven percent of Americans admit to stealing toilet paper from hotels.

Never-Ending Wipe Poo

Synonyms: *Stickum Stool, Double-Sided Deuce, Tar Turd, TP Thief*

While the delivery of this poo may go off without a hitch, the clean-up is another story. Eager to give your creation a quick glance, wipe, and be on with your day, this poo gets you hung up during the cleanup. The first signs of an impending problem appear during the first wipe when the toilet paper reveals a stickier–than–usual substance in a greater–than–expected quantity. The Never-Ending Wipe Poo (NEWP) has been likened to dark maple syrup because of its typical dark color and viscous consistency. When repeated wipes continue to yield significant amounts of residual stool, you wonder if and when you will ever finish the cleanup. Frustrated, you may modify the wiping technique by increasing the force applied or by employing a bidirectional, back-and-forth maneuver. Alas, the problem may not be with the wiping technique or the quality of the toilet paper, but in the unique attributes of the poo.

 Dr. Stool says: The Never-Ending Wipe Poo can be caused by one of several factors. First, the stickier the poo, the more you will have to wipe. Usually by the time stool makes it to your rectum it is somewhat desiccated by the colon's avid absorption of water. However, if your stool passes quickly through the colon, there will be less time for absorption and your stool may have a more gel-like consistency.

High-viscosity stool can also occur when there is bleeding from stomach or small intestine. When blood mixes with stomach acid, it forms a black, tarry stool called *melena*. This mixture of blood and stool makes its way down the GI tract and produces a particularly sticky stool which resembles tar in color and consistency (its smell is also notably abhorrent). When a NEWP takes on this form, it could be a sign that blood loss is occurring from the stomach, most commonly due to a bleeding ulcer.

Another, thankfully more common, explanation for the NEWP is fecal contamination of large hemorrhoids. These engorged blood vessels may become soiled as stool passes through the anal canal and can be difficult to clean. This is especially true for large, prolapsing hemorrhoids which move in and out of the anal canal during a bowel movement.

A fourth cause of NEWP is the presence of a perianal fistula, an abnormal conduit between the inside of the intestines and the skin. This results in continuous passage of stool from the anal region, thereby mimicking the NEWP. This unfortunate condition is discussed under "Poobiquitous" (page 103).

CHAPTER 9:

Friends and Enemas

Though forcing gallons of water up your butt is not most people's idea of a good time, when you need everything cleaned out, an enema is the way to go (literally!)

Ritual Enemas

Although we typically perform them for concerns about health or vanity, other cultures have used enemas for ritual practices. Enemas were particularly important in Mayan rituals, where they were used to deliver hallucinogenic substances used in religious rites. Ritual enemas were often depicted on Mayan pottery. It certainly makes for a conversation piece.

Royal Flush

In pre-revolutionary France, enemas were used to cure a variety of ills and were even performed on a daily basis after dinner. Lore has it that Louis XIV was a fan of this practice, with some historians estimating that he received upwards of two thousand enemas in his lifetime. Thanks to King Louis's enthusiasm for the procedure, it became de rigueur for the well-heeled members of his court. Society ladies received them regularly, believing they were essential for a good complexion. The Duchess of Bourgogne once got an enema at a party, administered by a maid who climbed under her ball gown, while the duchess conversed with the king himself.

Baby's First Enema

It's hard enough to give your baby a bath—never mind an enema. But for the Beng people of the Ivory Coast, it's a twice-daily ritual that begins as soon as the umbilical cord stump falls off. Beng mothers return to work shortly after birth, leaving their children with other caretakers, and need their infants to be bowel-trained as soon as possible. Giving the babies twice-daily enemas trains them to poop on command (at bathtime), freeing the mothers to return to the fields. It seems to work for them, but a warning: unless you are Beng, do not try this at home.

The Life Raft

Synonyms: *The Safety Net, The Trifecta, The Poo Pillow*

The mysterious practice of women traveling to the bathroom in herds has resulted in one major advance for mankind: the Life Raft. The Life Raft arose, in large part, because of the need for women to surreptitiously move their bowels in a public bathroom. Traditional tactics to mask droppage proved insufficient when in the proximity of five of your closest friends. Out of this necessity, the Life Raft was created—an ingenious innovation involving layering toilet paper in the toilet bowl prior to having a bowel movement in order to mask its sound. In fact, a properly executed Life Raft deployment results in "The Trifecta"—no noise, no splash, and no skidmark.

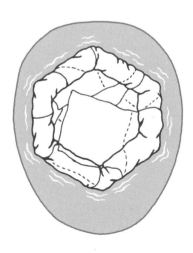

Dr. Stool says: The fear of backsplash has been around ever since people began dropping their deuces into a bowl of water. Small, dense particles of poo are at highest risk for causing a splash; this process is classically referred to as "buckshot." In reality, backsplash contamination, even from the grimiest of public toilets, is unlikely to cause serious health consequences. The Life Raft can help conceal poo's landing, but the only surefire way to prevent a splash of contaminated toilet water is to ensure the delivery of a single, lengthy bolus which effortlessly slides into place at the bottom of the toilet bowl. So to avoid embarrassing buckshot in front of your friends, eat lots of fiber and stay well hydrated!

CHAPTER 10:

Looking Backwards: Great Moments in Poo History

Though it's not often discussed, poo has played major roles in the world's development. Herewith, some highlights from its history.

Evolutionary Poo

Some scientists have credited poo with causing the explosion of complex organisms that occurred in the Cambrian period over five hundred million years ago. They claim that poo increased oxygen levels in the atmosphere, thereby allowing for the growth of more complex organisms. For some reason, Darwin never really focused on the critical role poop played in the evolution of species.

The study of poo not only gives us valuable information about our current digestive health, but it also can yield valuable information about the behaviors of the animals that roamed the Earth millions of years ago. The study of coprolites, or fossilized poo, is widely used by paleontologists to better understand the dietary habits of dinosaurs. The largest dino poo ever excavated was found in Washington, and measured 3 feet 4 inches in length. Not impressed? Keep in mind that the T. rex turd has undergone considerable shrinkage (65 million years' worth).

Prehistoric Poo

How old is the oldest surviving human feces? An astounding 50,000 years. Found in El Salt, an archeological site in Spain, the artifact provides valuable clues into the Neanderthal diet and lifestyle. Researchers were particularly surprised by evidence that suggests the Neanderthals ate plants as well as meat.

When in Rome . . .

In 2013, archaeologists excavating Nero's Roman palace found something rather extraordinary: a 50-seat bathroom. The toilets consist of side-by-side holes in a very long stone bench, without any walls or dividers for privacy. Researchers believe this was the restroom designated for use by slaves. Communal public restrooms were used by higher social classes as well. One of the less civilized features was their communal wiping-sticks. After use, the stick-mounted sponges would be placed in salt water to await the next customer. Yuckusmaximus!

●●

Nuggets: *Ancient Egyptian tombs had special toilet chambers for the pharaohs to use on their way to the afterlife.*

There Be Dragons

Wonder where the poop went in medieval castles? In some cases, right into the moat. Some castles featured simple toilets, consisting of a stone seat over a chute that emptied into the moat. A poop-filled moat would certainly ward off invaders . . .

Made in China

Chinese emperors in the fourteenth century were the first to use paper specifically for the purpose of toileting and ordered it in sheets that measured two feet by three feet. This seems luxurious by today's standards, where a typical sheet measures a measly 4½ inches square.

Gong Farmers

Hundreds of years ago, someone had to collect human waste from London cesspits and transport it outside city limits. These men were referred to as "gong farmers" because they supposedly cultivated "night soil," a reference to the requirement that fecal material only be transported in the dark. We're not sure where the "gong" part fits in. They were also known as "night soil men."

Although well compensated, the work was extremely unpleasant and very, very dangerous. The poorly ventilated cesspits, full of fecal fumes, could be a deathtrap. Nightsoil men were sometimes asphyxiated, or even burned, when the fumes ignited the lanterns they used to light their way.

Because it was so expensive to have the nightsoil—that is, poop—cleaned out of the local cesspit, cesspits in the poorer parts of eighteenth-century London rarely got emptied at all. Records show that one, which held the droppings from 176 local residents, was not cleaned out for fifteen years. Needless to say, overflowings were common—and highly odorous.

Poof

"Fart Proudly" is the title of a 1781 essay on flatulence written by Benjamin Franklin. In the text, Benjamin Franklin proposes, tongue in cheek, that the Royal Academy should use science to convert farts into something more agreeable to society.

Where No Man Has Gone Before

On earlier space missions trying to poop was a time-consuming and sometimes messy endeavor, requiring astronauts to strip down and poop into a "fecal bag" while other crew members tried not to watch in a very small space. It could take up to forty-five minutes, and poop sometimes escaped and floated around the ship. In one incident recorded on the Apollo 10 mission, astronauts Gene Cernan, Tom Stafford, and John Young debated who was responsible for a loose log wafting through the cabin.

Déjà Poo

Synonyms: *Veggie Burger, Leftovers, Corn-Backed Rattler, Sloppy Seconds*

"Haven't I seen that somewhere before?" Most notoriously involving corn, Déjà Poo is perhaps the most renowned and befuddling of all poos. A Déjà Poo is a bowel movement that has remarkably familiar portions of a recent meal embedded in it. This poo can include a potpourri of colors, often containing pieces of vegetables and other items that look as though they do not belong among the mass of poo in which they are entrenched. When producing this kind of turd, you may wonder whether you chewed sufficiently or whether your body extracted any of the nutrients from the food you just ate. You may also wonder how your body can process heavy meats and pastas but not an innocuous kernel of corn.

 Dr. Stool says: This "super-natural" experience is most often the result of consuming a meal loaded with insoluble fiber. While soluble fiber found in foods such as beans, nuts, and carrots forms a gel-like substance when mixed with stomach secretions, the insoluble fiber contained in oat bran (and yes, corn on the cob) passes through the GI tract largely unchanged. Humans lack the necessary enzymes to digest certain components of plant cell walls. The presence of these indigestible remnants embedded in your feces is what gives rise to the sensation of Déjà Poo. Consumption of high-fiber foods like corn and celery can soften the stool, thus yielding just as much enjoyment on the way out as on the way in.

Rotten Poo

Synonyms: *Napalm, Rancid Poo, Aftershock Poo, Agent Orange*

This poo can vary in shape and size, but its distinguishing feature is its atrocious and unbearable odor. As this poo is under way, the stench will overwhelm you. Even with a quick-response courtesy flush, survival instincts force you to speed up the defecating process in order to exit the bathroom as quickly as humanly possible. Lord help the innocent bystanders if you are in a public restroom, because this odor will linger and may promptly cause you and others to experience severe gagging and nausea. Worse than a rotten egg, worse than Limburger cheese, this poo smells as if a dead animal has been decomposing in your intestines and is making its exit at its most noxious moment. A Rotten Poo's odor is so powerful that anyone entering the vicinity within the next several hours is affected. There really is no way to prepare for a poo like this. However, when it happens, a quick termination of the stooling session is a must.

 Dr. Stool says: The compounds that lend poo (and farts) their odors are produced by the bacteria residing in our colon. These bacteria react with our ingested food to form smelly sulfur-containing compounds such as hydrogen sulfide and mercaptans. Occasionally, foul-smelling stools can be a sign of disease. Difficulties with digesting and absorbing food, as occurs in cystic fibrosis or chronic pancreatitis, can result in floating, greasy, and foul-smelling stool (see "Floaters vs. Sinkers," page 94). Intestinal infections, particularly with the parasite *Giardia lamblia*, can also cause diarrhea with an especially abhorrent odor. This infection is usually contracted after taking a dip in a freshwater lake or pond. Ingestion of even a small amount of contaminated water can be followed by several days of rancid diarrhea.

While persistent foul-smelling stool can be a sign of underlying disease, the occasional rotten-smelling poo should not be a cause for health concern. Your reputation at work, however, is a different story. If you deposit a Rotten Poo at work, Dr. Stool highly recommends leaving the stall of stench as quickly and stealthily as possible and taking a long walk outside while praying mightily that the smell hasn't permanently embedded itself in your clothes.

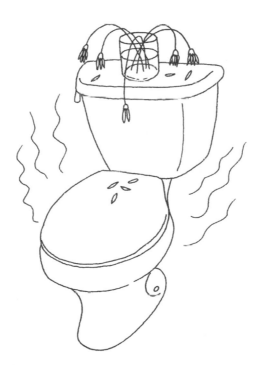

The Royal Wee

Everyone poops—even princes and presidents. Here are some surprising facts about our leaders' bathroom habits.

The King on His Throne

In the late 1600s, King Louis XIV of France regularly held official meetings while sitting on his beloved "throne." Known for his absolute command of power, Louis was an impartial ruler when it came to poo, unabashedly relieving himself in front of royalty and peons alike. Perhaps Louis XIV's comfort level with defecation contributed to his record-setting seventy-two-year reign of power. After all, what opposing ruler could effectively negotiate with the king when faced with the constant threat of having to witness his majesty's next bowel movement?

A Royal Mess

Although Versailles featured one hundred bathrooms and three hundred commodes, the residents, it seems, often preferred to do their business elsewhere, particularly on the staircases and in other public areas. By 1715 it had become so bad that officials agreed to have the hallways cleaned of poop weekly, and issued an announcement to that effect.

Right-Hand Man

Because English kings could not be expected to wipe their own bottoms, they were served by a "groom of the stool." The groom's duties included attending the king at the toilet and disposing of the royal remains. Although it sounds like a terrible job, it was actually very prestigious, as it granted close access to the monarch. It was usually given to courtiers of high rank, who were richly rewarded for their loyal service. The position wasn't discontinued until King Edward VII assumed the throne in 1901.

DYING ON THE THRONE

A surprising number of royals have died on the toilet. They include:

- Elagabalus, Emperor of Rome, stabbed while in the latrine

- James I, King of England, murdered while trying to escape from his bathroom into the sewers

- George II, King of Great Britain, suffered aortic dissection while sitting on the toilet

- Henry III, King of France, murdered leaving toilet

- Wenceslaus III, King of Bohemia, stabbed in the privy

- and, of course, the King himself, Elvis Presley, who died of a heart attack while sitting on the toilet

●●●●●●●●●●●●●●●●●●●●●●●●●●●●●●●●●●●●●●●

Nuggets: *A number of royals died on the toilet, and at least one was born on it: Holy Roman Emperor Charles V. Charles who was delivered by his mother, Joanna the Mad, into a filthy chamber pot in 1500.*

Founding Flusher

Given that Thomas Jefferson was an enthusiastic innovator, it should come as no surprise that he was one of the first Americans to own a flush toilet. He installed three of them in the White House. Even more innovative? They ran on rainwater, collected in cisterns in the attic.

Stinkin' Lincoln

Lincoln is known for his noble achievements, but in his day, he was known for something else entirely: his love of potty jokes. Lincoln was reportedly a treasure trove of jokes and humorous stories, and favored the scatological ones. Unfortunately, no record of his favorite jokes survives, though it's said he once made a court bailiff laugh so hard he fell out of his chair.

Very Secret Service

On a trip to Vienna, the White House flew in a special presidential crapper so that President George W. Bush's feces could be collected and disposed of in a secure manner. Secret Service agents capture Presidential Poo in order to prevent foreign intelligence agencies from collecting information about the commander in chief's health. Governmental agencies, including the United States' C.I.A. and the Israeli Mossad, have used this approach to gain valuable information on the health status of world leaders such as Mikhail Gorbachev and former Syrian President Hafez al-Assad.

Supremely Constipated?

Among the strangest claims North Korea made about its late Supreme Leader Kim Jong-il? He was the greatest golfer in the world; he invented the hamburger; and he did not pee or poop. According to North Korean textbooks, Jong-il was a supernatural being who transcended normal human biological processes.

CHAPTER 11:
Potty Talk

From the loo to the lavatory, from dung to dookie, we've got lots of names for the restroom and the things we make in there.

The Scoop on Poop

According to Eric Partridge in his book *Origins: A Short Etymological Dictionary of Modern English*, "poop" comes from the Middle English word poupen or popen, an onomatopoieic word that originally meant "fart." Robert Chapman, author of *American Slang*, writes that "poop" came into use with its current meaning around 1900. As for poop deck, that's got nothing to do with pooping at all. It's believed to be derived from the Latin word puppis, which refers to the ship's stern. This is probably where we get the phrase "pooped out"—to be pooped, on a ship, was to be overtaken by a wave from behind, where the poop deck is located.

The word "feces," first used in the fifteenth century, comes from the Latin word *faeces*, meaning sediment or dregs. It was being used to describe human poo by the 1630s. As for "crap," it probably derives from the Middle English word *crappe*, meaning grain chaff. By 1846 it had acquired the meaning it has today. Contrary to popular opinion, its origins have nothing to do with toilet pioneer Thomas Crapper, whose name is just an extraordinary coincidence.

The origin of "number two" is more mysterious. Although the term isn't that old, it first appears in print in 1902. It's possible that it comes from rhyming slang (two=poo). It may also refer to the fact that people typically pee before they poo.

●●

Nuggets: *Why do colon (as in "intestine") and colon (as in ":") sound the same? Because they have the same Latin root. Both come from the Latin word for "limb" or "member."*

Poppycock, Poop Twigs, and Pastrami on Devil's Fart

Lots of things that have nothing to do with bathroom matters (we hope) were named for things that do. "Poppycock," for instance, has come to mean "nonsense," but its original meaning was a little more specific. It comes from the Dutch *pappe* (soft) *kak* (poop)—in other words, diarrhea.

And before you pucker up under the mistletoe this year, you may want to know that "mistletoe" literally means "poop twig." Mistletoe seeds are transported by birds, who excrete them in their poop. Ancient Anglo-Saxons noticed that the plant seemed to grow from bird-droppings in the trees, and gave it the name *missel* ("poop") *tan*, or *toe* ("twig").

Pumpernickel bread is delicious, but it might not taste so good once you know what the name means. It comes from old German words meaning "devil's fart," and is apparently a reference to the bread's after effects.

HOW TO SAY POO AROUND THE WORLD

- Arabic: khara
- Chinese: pihuà
- Czech: hovno
- Danish: skid
- Dutch: drol
- Farsi: ann
- Finnish: paska
- French: merde
- German: scheisse
- Greek: skata
- Hebrew: kaki
- Hungarian: végbélsár
- Italian: merda
- Japanese: kuso
- Korean: sheeba
- Latin: stercus
- Polish: filmu
- Portuguese: merda
- Russian: govno
- Spanish: mierda
- Tagalog: tae
- Thai: kee
- Turkish: batmak, bok

Braille Poo

Synonyms: *Baby Ruth, Porcupine, Rocky Road*

Despite the fact that all poo follows the same route through the GI tract before coming to rest in the toilet bowl, individual bowel movements can have markedly different textures. While some bowel movements are smooth and silky, others take on a more angular and bumpy appearance. Braille Poo can be identified by its rough and uneven texture. The passing of these uneven poos makes you wonder what causes this type of potholed feces and why they all can't achieve the honeyed smoothness of previous poos. However, it is important to accept that just as some roads are well paved while others are gravel or cobblestone, poo's texture can similarly be quite varied.

Dr. Stool says: This amalgam of poo appears to be due to a "catch-up" phenomenon. The slowing of colonic transit (i.e., constipation) allows digestive debris from several meals to form a single, variegated bolus of stool. In other words, the ham-and-cheese sandwich you had for lunch "runs into" the scrambled eggs you had for breakfast, the steak you had for dinner the night before, etc. Identifying remnants of prior meals can be both challenging and enjoyable. Unlike Déjà Poo (page 81), where the food item is identified largely unchanged in the turd, Braille Poo requires careful scrutiny of the stool's color, texture, and occasionally odor in order to correctly identify the individual components.

Floaters vs. Sinkers

Synonyms: *Aircraft Carrier vs. Submarine, The Buoy vs. The Anchor*

One of the most mystifying characteristics of poo is the tendency of some turds to float and others to sink to the bottom of the bowl. Whether big or small, brown or black, it is impossible to predict whether a poo will be a Floater or a Sinker until it hits the water and settles in. The obvious benefit of a Floater is that it won't leave racing stripes (see page 25) in the bowl. However, some stubborn Floaters have been known to resurface after multiple flushes, a distant cousin of the Déjà Poo (page 81).

 Dr. Stool says: There are two components of stool that cause it to hover on the surface of the toilet water: gas and fat. Most commonly, Floaters are due to the fourth burrito or second helping of chili from the day before. When gas is the culprit, you may also notice your fart frequency increasing above normal levels. If the Floaters last for a day or so, there is no reason to be alarmed.

Foul-smelling (really foul!), greasy, floating stool is more worrisome. It usually indicates the presence of fat in the stool. This is never normal and often reflects an underlying problem in the GI tract, most commonly involving the pancreas or liver. The pancreas, liver, and gall bladder normally team up to help the body digest the fat we consume. When these organs become diseased, dietary fat passes through our GI tract largely undigested and results in the formation of floating, "oil-slick" stool.

CHAPTER 12:

Poo Hall of Fame

These folks made this world a better place to pee, poop, and fart in.

Claudius

While most modern cultures shun those who pass flatus in public, a forward-thinking ruler, Roman emperor Claudius (10 BCE–54 CE), first legalized farting thousands of years ago out of concern for his subjects' well-being. By lifting the ban on farting at banquets, he was in line with the prevailing notion that retaining flatus could be harmful to one's health (not to mention painful!).

Sir John Harrington

Author. Royal. Toilet inventor. Four hundred years ago, Sir John Harrington may have been the most interesting man in the world. A favorite courtier of Queen Elizabeth, Harrington was known for his poetry, but he also produced something far more important: the first flush toilet. He was a few centuries ahead of indoor plumbing, however, so the idea didn't really go anywhere. But Harrington did leave the world another gift: he's the ancestor of Kit Harrington, *Game of Thrones'* handsome Jon Snow.

Henry Moule

In the nineteenth century, Henry Moule, an English vicar, invented a new kind of toilet: the earth closet. It functioned not unlike the modern toilet, but instead of dousing the waste with water, it dumped soil on top of it. Earth closets were popular for a time, especially in the countryside, but were eventually replaced by the modern flush toilet.

Thomas Crapper

Perhaps because of his name, nineteenth-century British plumber Thomas Crapper is often credited, wrongly, as the inventor of the first flush toilet. But what Crapper did invent was perhaps even more important: the first reliable flush toilet. It featured a raised tank and a pull chain, and was sold under the name the Marlboro Silent Water Waste Preventer. Crapper's machine promised—and delivered—odor-free waste removal, and made him a very wealthy man.

Joseph Pujol

In the nineteenth century Joseph Pujol became famous as a "fartiste," a virtuoso of flatulence who performed before large, appreciative audiences. Better known by his stage name, "Le Pétomane," or "fart maniac," Pujol had the remarkable talent of being able to fart at will. His tricks included playing the flute with his anus and farting to blow out candles stationed several yards away. He also had the ability to recreate animal sounds and typically opened acts with his own, very special, rendition of the French national anthem. But it seems like a lost opportunity that he was French and not Spanish; had he been the latter, his name would have been pronounced, quite fittingly, "poo-hole."

Monster Poo

Synonyms: *Lincoln Log, The Crowd-Pleaser, Double Deuce, The Five-Minute Diet*

You may wonder, "How did something that large come out of me?" While sitting on the toilet and vigorously straining to discharge a poo of this size, you feel like the turd took a wrong turn in your intestines and is attempting to come out sideways. You may feel the swelling of veins in your forehead and the beading of perspiration as you toil to force this mass of poo out of your system. Despite the strain, this internal bodily struggle will continue until the last of the turd exits. After discharging this Double Deuce, it is not uncommon to feel as though you just lost five pounds. For a quick second, you may even consider summoning your friends to witness firsthand the greatness of your feat.

Despite the fact that these poos are not always the easiest to discharge, there is a great feeling of accomplishment and pride associated with the deposit of a Monster Poo. In addition to its massive girth, Monster Poo's most characteristic feature is its tendency to extend beyond the water surface. You may even fear flushing it without first using the toilet brush to break it up into smaller pieces. Despite the separation anxiety that may result, we recommend promptly flushing the toilet— after you have had a chance to bask in the glory of your poo.

Dr. Stool says: Although studies have not correlated the degree of straining with the size of the bowel movement, several factors play a role in creating a glacier-size turd. The "bulk" of the stool is directly related to the amount of fiber and water you consume. Picture the engorged appearance of your favorite legume after soaking it overnight in a bowl of water! A similar reaction takes place in the gut, where soluble fiber and water combine to form a swollen mass of turd.

CHAPTER 13:

It's a Smell World After All

Everybody poops, but we don't all do it the same way. Here's how it goes down around the globe.

Left Out

An estimated 70 to 75 percent of world's population—nearly five billion people—don't use toilet paper. What do they use instead? In some countries, bidets; in others, their hands. For this reason, it's considered rude in many countries to eat or touch others with your left hand, as the left hand is reserved for wiping.

Faux Pas

It is not uncommon for public and private bathrooms in Japan to come equipped with toilet slippers. This fecal footwear is for bathroom use only and is yet another example of the Japanese's strict attention to hygiene. Much to the chagrin of their hosts, visitors have often mistakenly worn these slippers, which can be quite plush, freely throughout the home.

DOO YOU KNOW?
POO AROUND THE WORLD

The amount of stool expelled per day varies from country to country. The average weight of stool in England is 106 grams per day, while South Asians unload nearly four times that amount. This difference is largely due to the higher fiber content in the average Indian diet.

El Caganer

Visitors to Catalonia, Spain, are often surprised to find a pooping figurine among the nativity scene. Known as El Caganer (Catalan for "the pooper"), these figurines date back to the seventeenth century and feature a red-capped, squatting peasant dropping trou. Its origins are unclear, but it may be a reference to fertility, or simply a harmless joke. The figurines are also seen in Andorra and parts of Portugal, Italy, and Spain.

Good Luck, Indeed

In Polish, the word fart means "good luck." For this reason there are a number of Polish products that probably wouldn't sell so well in the English-speaking world, including a "Fart Bar" candy and the even more disturbingly named "Fart Juice."

Coffee, Tea, or Pee?

In Ghana, English-speaking visitors are often startled by the name of a popular local soft drink: Pee Cola. The beverage does not, in fact, contain urine; "pee" means "very good." It's also a common Ghanaian surname.

Destination Poo

Turns out you can visit "poo" on three different continents: Fernando Poo, now known as the island of Bioko, off the coast of Africa; Poo Pathi, a holy site in southern India; and a Spanish district known simply as "Poo."

Poobiquitous

Synonyms: *Perpetual Poo, Poofinity, Stool Seepage*

A couple of loose, watery bowel movements after bad fast food is one thing; continuous seepage of stool from your rectal area is another. Despite moving their bowels several times a day, people suffering from Poobiquitous frequently discover their undergarments soiled with stool. The initial reaction upon finding yourself in this situation is to assume a hasty cleanup is to blame. When adoption of a more aggressive wiping protocol fails to stem the tide of liquid stool emerging from your rectum, you resort to placing wads of toilet paper in your underwear, hoping that this poo-in-perpetuity will soon cease. Ultimately, you question: Why the sudden, ceaseless seepage and how can it be stopped?

Dr. Stool says: Poobiquitous indicates the presence of a perianal fistula, an abnormal connection between the intestines and the skin surrounding the anus. In laymen's terms, this means that stool is coming out of your skin. Normally, stool is held in our rectum by contraction of the anal sphincter, a muscular valve that allows us to hold onto stool until a convenient time and place. Inflammation in the bowel, most commonly due to Crohn's disease, can cause the intestinal tract to erode into the perianal skin. This passageway allows stool to bypass the anal sphincter and results in continuous and untimely stool seepage. Historically, fistula treatment required surgery, but aggressive medical therapy with immunosuppressive drugs can now lead to healing in a majority of patients.

Ring of Fire

Synonyms: *Acid Poo, Hooters Souvenir, Curry in a Hurry, Fire in the Hole, Tabasco Turd, Feeling the Burn*

Sometimes, you sit down to do your business and a burning sensation rips through your anus. As every millimeter of poo passes through you, the burning only gets more excruciating. It may feel as though someone is funneling hydrochloric acid through your sphincter, and you may scream out, "Why?!?!?" As you pray for the poo to end and the burning to dissipate, you think back on what you have done to deserve this agony.

 Dr. Stool says: This sensation is most often due to consumption of spicy foods. Spices such as chiles and cayenne pepper cause direct irritation to the lining of the gastrointestinal tract. The exquisite pain felt upon defecating should come as no surprise because a similar burning sensation was surely experienced in the mouth during consumption of the spicy food. The similarity in sensation can be attributed to the fact that the inner layer of the mouth and the anus are lined by the same type of cell. These "squamous" cells, unlike the "columnar" cells lining the majority of the gastrointestinal tract, are able to discriminate among multiple stimuli, none more dramatic than its perception of five-alarm chili.

The Chinese Star

Synonyms: *The Dorito, Iceberg, Glass Shard, Mystery Poo*

This poo's defining characteristic is the excruciatingly painful sensation of feeling as if your rectum is being torn apart from the inside as the turd exits your body. This searing agony is commonly the result of passing a particularly hard, angular bowel movement. At times, this stool's appearance can be a source of bewilderment. As we hold back the flow of tears, our  awareness quickly shifts to the identification of the offending particle. The mounting rage, however, is diffused when we gaze into the waters only to see a small, seemingly innocuous turd resting peacefully on the bottom of the toilet bowl.

 Dr. Stool says: If the intense rectal discomfort persists despite elimination of all glass-containing items from your diet, you may have an anal fissure. An anal fissure is a tear in the lining of the anal sphincter, usually occurring after passing a particularly hard stool. This break in the lining causes spasm of the internal anal sphincter (similar to any other muscle cramp) and can make having a bowel movement feel as if you are passing razor blades. Treatment consists of topical anesthetics, stool-softening agents, and sitz baths.

CHAPTER 14:

Poo Zoo

We may have the power of speech and opposable thumbs, but
animals have some pretty great poops.

Does a Bear Poop in the Woods?

Not in the winter. During hibernation, bears continue to produce a small amount of feces, but they don't expel it. Instead, the feces forms a plug that will remain in place during the months of rest. Those who've had occasion to smell the fecal plug report that it smells mild and not unpleasant. But what about pee? They don't do that either, even though they're still producing urea and other waste products. Normally, a buildup of urea would be dangerous, but during hibernation bears go into a sort of recycling mode that lets them break down the nitrogen in urea and convert it to protein. That way, they preserve muscle mass during their long nap, and don't die of toxic buildup besides.

The Circle of Life

In the animal world, poo is not just natural—it's delicious. Lots of species eat poo, including dogs, gorillas, many insects, rabbits, and other rodents. Rodents do it because the plants that form the basis of their diet are hard to digest, and they can't get all the nutrition out the first time around. (Lucky for them, the poop in question smells better than most. Both guinea pigs and rabbits can produce poop that smells sweet and candy-like.) Other animals eat poo because it contains nutrients produced by intestinal bacteria. And still others— such as dogs—eat it as a source of protein.

Several animals eat their own poop, but none do it as enthusiastically as capybara do: they've been known to eat their poop as they're pooping it. Like rabbits and guinea pigs, they need to eat poop to get the nutrients they didn't absorb on its first past through. And they apparently find it so tasty they'll suck it out of their own rectum.

KNOW YOUR DROPPINGS

There are specific names for many animal droppings.
A sampling:

Animal	Droppings
Bear, boar	Lesses
Cows	Bodewash, tath
Deer	Fumets
Dogs	Scumber
Earthworms	Worm cast
Fox	Billitting
Hawks	Mutes
Otters	Spraints
Rabbits	Crotiles, crotisings
Seabirds, bats	Guano
Sheep	Buttons
Vermin	Fuants

Dung-Showering

Some animals communicate with sounds, some with gestures. Hippos, however, communicate with poo. Called dung-showering, the practice consists of dumping massive amounts of poo and pee, which the hippo then sprays around with its tail. It's done to scare off predators and mark territory, and sometimes to indicate submission by females.

Splendor in the Frass

Caterpillar poop is called "frass," and it has a number of uses. Gardeners prize it as a good source of nutrients for plants and soil. As for the caterpillars themselves, they use it to make a sort of umbrella, to keep themselves dry when it rains.

Wild Things

Pretty much all house cats bury their poop, but in the wild, only unassertive cats do. Animal behaviorists say this is a sign that house cats view themselves as submissive to their owners. If your cat poops openly—say, on your bed—it might not hold you in a whole lot of esteem.

Powerful Penguin Poop

They can't fly, but penguins can poop. Penguins poop with four times the force that humans do (this allows them to expel their poop a good distance from their nests). And because their poop is dark, and their snowy home is light, the poop can actually be seen from space. This, in fact, is the way scientists monitor penguin population: by studying their poop formations via satellite. Now that's some powerful poop!

Weird Wombats

Wombat poop has a shape unlike any other in the animal kingdom: it's cubical. The strange shape is a result of the poop's extraordinarily dry texture. When it emerges, it breaks off at more or less right angles, resulting in the distinctive cubical shape. Wombat poop is the driest in the animal kingdom, due to the wombat's extraordinarily slow digestive process—it takes 14 to 18 days to produce a poop. The slowness doesn't seem to effect output, however; wombats still manage to produce about 100 of the oddly cube-shaped poop pellets daily, which they use to help them find their way home.

Pooped Out

Given their reputation for, well, slothfulness, it's no surprise that sloths can't be bothered to poop very often. In fact, they typically only poop once a week. When they do, it's an event: they climb down from their jungle perch, grasp on to the base of the tree, and do a dance that watchers report to be highly entertaining.

A Whale of a Poop

True to their name, blue whales have colorful poop, but the color depends on what they've been eating. Eating krill produces orange poop; eating shrimp produces pink. The blue whale's poo is so large it can be seen from an airplane.

The Snake

Synonyms: *Curly Fry, The Long Thin Line, Fettuccine Feces*

The Snake is a thin and windy defecation that can contort itself into a variety of different shapes and sizes. Some Snakes wrap around the bottom of the toilet bowl clockwise, and others go counterclockwise. Some Snakes twist themselves into the shape of a pretzel, while others zig-zag across the bowl. Regardless of the final form, Snakes seem

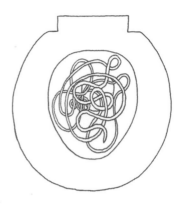

to go on forever as they leave your rectum. In the end, however, Snakes never leave you with the same feeling of accomplishment as do poos of a wider diameter.

Dr. Stool says: Excessive straining usually forms this long, thin stool. The act of bearing down causes contraction of the external anal sphincter (the "valve" that opens in order to allow feces to exit the rectum). Contraction of this muscle narrows the aperture through which the stool bolus passes, thus creating your garter-snake-morphed turd. While everyone will occasionally produce these slender stools, progressive narrowing in stool caliber over months can indicate the presence of a rectal cancer. These "pencil-thin" stools are formed as the rectal tumor's growth gradually narrows the colonic cavity.

DOO YOU KNOW?

SPAGHETTI "POO"

When is a poo not really a poo? The passage of this noodle-like strand may at first seem to be a particularly slender thread of stool. Closer inspection, however, will reveal that this poo imitator is, in fact, a parasite known as Ascaris lumbricoides. These worms, somewhat reminiscent of angel hair pasta, can grow to be over a foot long. They reside quietly for years in the small intestine and may only come to attention during their dramatic exit. Rarely, these worms can cause nutritional deficiency by competing with your GI tract for valuable nutrients. Think you have a worm? You're probably not alone; research has shown that one-quarter of the world's population is infected with this roundworm.

CHAPTER 14:
Artsy Fartsy

From Mozart to Marcel Duchamp, bathroom matters have inspired artists for centuries.

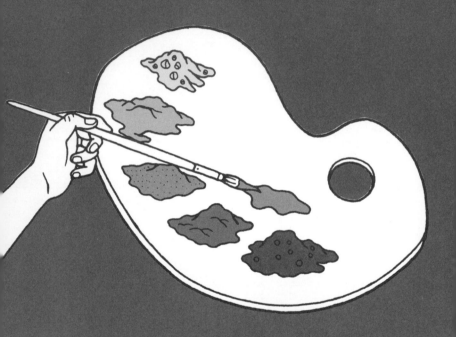

L'Art de Chier

Everything sounds classier in French. This includes the 1806 poem by Argaud Debarges, *L'Art de Chier*. In English, this translates to "The Art of Taking a Dump," and this is exactly what the poem describes. Debarges helpfully suggested readers wipe with it once they'd finished reading.

Worth Its Weight in 40-Year-Old Poop

In 1961 Italian artist Piero Manzoni was inspired to fill 90 cans with his own feces, label them "Artist Sh*t," and price them according to their weight in gold. Forty years later, they're actually worth much more. Over the years, as gases from the decomposing poop built up inside the cans, many of them exploded. The remaining few became worth a good deal, and can fetch well over $100,000 each.

DOO YOU KNOW?

POOP THE RAINBOW

Want to custom-color your poo? You can, without ever resorting to food coloring (which will also do the job). Spinach and blueberries can turn it black; beets can make it red; blackberries can make it green; cantaloupes and sweet potatoes can make it orange; and milk can make it yellowish.

The following chart can help explain the variations in poo hue.

Black:

Iron pills (or foods high in iron), bismuth compounds (i.e., Pepto-Bismol), blood (from higher up in digestive tract)

Red:

Blood (from lower down in digestive tract), beets

White/Gray:

Bile duct blockage, liver disease

Green:

Gastrointestinal infection, spinach

Yellow:

Fat in stool (pancreatic disease)

●●♦ ●♦●♦●♦ ●♦●♦●♦ ●♦ ●♦●♦● ●♦●♦●♦●♦●♦● ●♦●♦●♦ ●♦

Nuggets: *Famed architect Le Corbusier called the toilet*
"one of the most beautiful objects industry has produced."

An International Stink

In September 2008, a giant inflatable feces sculpture by artist
Paul McCarthy broke free from its moorings at the Paul Klee
modern art museum in Switzerland before bringing down
a power line. The gigantic turd eventually came to rest 200
yards away in the grounds of a children's home, but not before
causing an international stink about what constitutes art.

Sixty-Foot Feces

In the fall of 2008, controversial artist Andres Serrano held
an exhibition in New York City in which he featured photos of
human and animal excreta, including a sixty-foot image of his
own bowel movement, entitled "Self-Portrait." Nobody loves
poo as much as we do, but 60 feet?

National Poo Museum

If you love poo, and you love museums, then you'll be
crazy about the National Poo Museum on the UK's Isle of
Wight. The museum features a wide range of poop samples
(thankfully enclosed in resin) and interactive poo-related
exhibits. Its mission is "to lift the lid on the secret world
of poo—to examine our relationship with it and to change
forever the way we think about this amazing substance." The
museum has a gift shop with an online store, and they ship
abroad.

Camouflage Poo

Synonyms: *Dalmatian Dookie, Chocolate Chip Cookie Doo, Black and Tan, Olive Loaf*

A reference to a multitoned poo (varying shades of black, brown, and green), Camouflage Poo looks like a mosaic of diverse excrement from different sources and from assorted meals. While it is all one unit of poo, it resembles a conglomerate rock, with many pieces of poo forced together by the impact of time and pressure. Due to their camouflage coloring, such poos have historically been difficult to identify in the wilderness. Thankfully, the present-day practice of using toilets affords us a pristine white backdrop against which we can carefully examine these poos. While the size of this type of turd varies, the texture is usually bumpy and the appearance unmistakable.

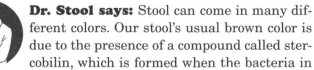 **Dr. Stool says:** Stool can come in many different colors. Our stool's usual brown color is due to the presence of a compound called stercobilin, which is formed when the bacteria in our colon digest bile. Most daily variation in stool color is due to the dietary intake of various foods and medications. However, changes in stool color that persist for longer periods of time can be a sign of an underlying gastrointestinal disorder.

The Green Goblin

Synonym: *Seaweed Stool*

The Green Goblin, though thankfully rare, is an explosion of foul-smelling diarrhea that is characterized by its viridian hue. Although various foods can lend normal poo a greenish tint, this bowel movement's deep-set, blackish green appearance makes it seem as if you are viewing it through night-vision goggles. Also distinguishing this poo from more benign green poos is its liquid form and associated symptoms of fever and abdominal pain. More often than not, this poo transformation occurs following a course of antibiotics for a tooth abscess or sinus infection.

Dr. Stool says: This diarrheal illness is caused by the overgrowth of a specific bacterium in the colon called Clostridium difficile. The use of the antibiotics upsets the natural balance of "good" and "bad" bacteria in our intestines and allows the proliferation of this particularly harmful organism, which, in turn, results in inflammation of the colonic lining and causes profuse diarrhea, abdominal pain, and fever. Medical treatment is with a different type of antibiotic that specifically attacks Clostridium difficile.

Rambo Poo

Synonyms: *Uh-Oh, Chocolate Sundae with Strawberry Sauce, The Neapolitan Poo*

On rare instances, you may look down at your droppings and see traces of blood. Despite the tone of this book (and the title of this poo), this is no laughing matter and could represent several different serious problems that you should consult a physician about.

Thankfully, there are many benign causes of blood in the toilet bowl, such as hemorrhoids, diverticulosis, and arteriovenous malformations (abnormal blood vessels that have a tendency to bleed). Before overreacting, keep in mind what happens when you put a few drops of food coloring into a bucket of water. Similarly, a few small drops of blood will convert your toilet into a large, unwelcome bowl of fruit punch.

Often overlooked, vigorous overwiping, which causes a little blood to appear on the ol' brown starfish, could also be the cause of the bloody surprise. In this case, consider yoga, exercise, or a new hobby for alternative stress relief. A conversion to a softer toilet paper would also be prudent.

However, the most feared cause of blood in the stool is colon cancer. Due to the seriousness of this illness, any new sighting of reddish stool or blood-tinged toilet water should always be followed by a visit to your doctor. In most cases of gastrointestinal bleeding, a colonoscopy is performed to visualize the interior of the GI tract and identify the source of bleeding.

Dunce Cap

Synonyms: *Conehead, Tapered Wafer, Biggie Small*

As this type of bowel movement begins, you strain and feel like you are on the precipice of another Monster Poo (page 98). However, when the cork is popped and the poo begins to leave your body, it becomes exponentially easier from beginning to end as the bowel movement flows out. After completion, when you gaze at your conquest, you notice that the poo began with a thick base but then scaled down to a point. While you feel refreshed now that you have finished defecating, you wonder whether the initial pain and strain were necessary. You long for a more uniform poo, where the bulk is more evenly distributed.

Plagued with one too many Coneheads? Try relaxing on the toilet. Resist that urge to contract your abdominal muscles in order to more quickly expel the log. Find a quiet, isolated stall where you won't feel pressured to quickly finish the deed. Still finding yourself tensing up at the wrong moment? May we recommend reading a good book about poo to help you "loosen up."

Dr. Stool says: Remember when your mother said, "Don't try to be something you're not"? You had no idea this sage advice would apply to your defecating practices. The tapering appearance arises from one's attempt to craft an otherwise modest amount of stool into a long, robust log (as in kindergarten, when you tried to roll a long snake out of a small piece of clay). The narrow ending is a result of your continued straining and the pinching motion of your external anal sphincter (like squeezing the last bit of icing out of a pastry bag).

Wipe Right

Partners need to be compatible not just in the bedroom, but also in the bathroom. How do you feel about farting in front of each other? Do you pee with the door closed or open? What about poop? So many things to consider before you swipe right.

Romantic Restroom

George Washington's estate, Mount Vernon, is best known for its dramatic two-story salon, its cupola, and colonnades. Less well-known, but no less remarkable, is its bathroom featuring two side-by-side commodes. We do not know if George and Martha ever used them for a romantic restroom rendezvous, but we can hope.

●●●●●●●●●●●●●●●●●●●●●●●●●●●●●●●●●●●●●

Nuggets: *Men pee faster than women until about age 50. After that, the women pull ahead.*

Bridal Shower

Today we celebrate a wedding by throwing rice or confetti. But in some North African weddings, it was considered good luck to throw the urine of the bride instead. Urine may also have played a role in ancient English and Irish weddings; some historians think the wine that's part of the ceremony today was a replacement for urine in an older tradition.

Drunk in Love

Like humans, giraffe courtship typically begins over a drink, but unlike humans, that drink is pee. The male giraffe starts things off by rubbing the female's behind until she pees. He then tastes a mouthful of the female's urine to determine whether or not she's in heat, and if she is, he pursues her. The process is known as the Flehman sequence.

Nuggets: *In seventeenth-century Germany, the accepted way for a young lady to end a romantic relationship was to put a bit of poop in her would-be suitor's shoe.*

L'Amour

French weddings end with a strange tradition. At some point during the wedding night, the wedding party hunts down the newlyweds and serves them a snack—in a chamber pot. The treat is usually something like chocolate-covered bananas or sausages floating in champagne and adorned with rice paper, and the interlopers won't leave until the bride and groom have taken a bite.

The Honeymoon's Over Poo

Synonym: *The "I Do" Doo*

Courting and dating rituals may vary across cultures and
time periods, but one constant is the anxiety involved in
taking a dump while in the presence of your significant other.
In the early stages of a relationship, many people come up
with elaborate stories to avoid the issue, often suppressing the
urge to poo for days at a time. Ultimately, as the relationship
progresses, you begin to openly proclaim your desire to
poo, eventually gaining the confidence to poo freely in that
person's presence. Allowing your significant other to smell
your poo without concern officially lets you know that the
honeymoon is over.

THE INTIMACY OF POO

Nature is full of examples suggesting that defecation is more than just physiologic necessity. One need look no further than at a cat hastily concealing its feces, a dog doing the reverse leg kick to separate himself from his deed, or a young child quietly escaping to a quiet corner of the room to appreciate the extremely personal nature of poo. Given the intimacy of this process, it comes as no surprise that there is an air of vulnerability surrounding this act. This sentiment is embodied in the well-known saying, "He was caught with his pants down." Some would say that allowing oneself to "be caught" by another while performing this most private of duties truly signifies the development of an unwavering trust. Others would say that pooing openly in front of others is just weird.

Dutch Oven

Synonym: *Bed Warmer*

When someone who is lying awake in bed farts under
the sheets, then pulls the sheets over the head of their
unsuspecting partner, it is called giving him or her a
Dutch Oven. While a Dutch Oven is often seen as a cruel,
juvenile act, openly farting in front of a partner signifies the
development of an unwavering trust (or over-familiarity) in
a relationship. The Dutch Oven is an extreme form of the
Honeymoon's Over Fart, a close cousin of the Honeymoon's
Over Poo. There is some debate as to whether the open pooing
or farting milestone comes first in the natural progression of a
relationship. One thing is certain, however: the honeymoon is
most certainly over for couples that exchange Dutch Ovens.

CHAPTER 15:
Mind Your Pees and Qs

Unless you're an astronaut, you don't pee in a vacuum. We poop, pee, and fart in a society with certain expectations. Those expectations, however, have varied.

What Would Miss Manners Say?

In the sixteenth century, etiquette manuals became quite popular, and they were a good deal more frank than etiquette guides today. Erasmus's 1530 manual advised, "Do not move back and forth on your chair. Doing so gives the impression of constantly breaking, or trying to break, wind."

Courtesy Flush

The Courtesy Flush is an act of flushing immediately upon defecation to minimize total odor exposure time. Most often employed in a public facility, this technique has also been used to disguise embarrassing sounds that may result, say, from a bout of explosive diarrhea.

Doo the Right Thing

Today we pick up dog poop because it's good manners, and, in most places, the law. But in Victorian England, people did it because it was profitable. Leather makers relied on the enzymes in feces to tan the leather, and children often earned money by selling them the dog poop they found on the streets.

Stage Fright

Stage fright has plagued mankind since we began peeing in the company of others. The typical stage fright scenario occurs at a ballpark or other places where herds of men pee in small confined spaces. Pee Anxiety (aka Tinkle Terror or Ballpark Bladder), also known as paruresis, or Shy Bladder Syndrome, is surprisingly common, with 7 to 10 percent of American men reporting difficulty urinating in close proximity to others. For a majority of paruresis suffers, the issue is psychological, a form of anxiety that typically begins in the teenage years and may stem from embarrassment at having the genital region exposed. In older men, alcoholic beverages can cause an already enlarged prostate gland to swell, causing obstruction to urine flowing out of the bladder.

●●

Nuggets: *Studies have shown that when someone is standing near them in a public urinal, men take longer to start peeing, but finish the job 40 percent faster.*

DOO YOU KNOW?

WHY DOES YOUR POO SMELL WORSE TO OTHERS?

This is one of the great mysteries of poo. Keeping track of our bowel movements is an important aspect of our everyday lives (why else would we be programmed to take that quick glance into the toilet after a bowel movement?). Surely, detection of poo's aroma would seem to be advantageous from an evolutionary perspective. For instance, serious ailments such as bleeding and infection can often produce changes in smell. Regardless, it is best to acknowledge the universal smelly nature of poo (even our own) and give a quick courtesy flush halfway through a bowel movement.

The Victimless Crime

Synonyms: *Odorless Fart; No Harm, No Foul; Without a Trace*

The opposite of the SBD (page 48) is the Victimless Crime. These odorless farts provide immediate physical relief from the gaseous pressure while the lack of aroma lends some equally important psychological relief. The Victimless Crime is like a delicious low-calorie dessert—all the pleasure with none of the guilt. Although highly desirable, especially for those working and living in close quarters, the odorless fart is also unpredictable. It should be noted that while claims of odorless flatus are plentiful, the truly odorless fart is indeed rare. Just as there are parents who believe their kid to be the most adorable on earth, there are equally delusional individuals who allege passage of unscented flatus as their norm.

 Dr. Stool says: Despite our efforts, newborn babies are the only humans who can truly perpetrate a victimless crime in the realm of flatulence. The newborn intestinal tract is devoid of bacteria (bacterial colonization begins immediately after birth) and therefore the majority of expelled air is composed of nitrogen and carbon dioxide, both odorless gases. Theoretically, aroma-free farts could be achieved by dietary elimination of all sulfide-producing foods and eradication of all intestinal bacteria through antibiotics or colonic "cleansing." As desirable as odorless flatulence may be, none of these strategies are advisable.

Pungent Pee

Synonyms: *Smelly Stream, Asparagus Aftershock, Scrambled Eggs*

In most instances, freshly produced pee is remarkably odorless. (A pee puddle that has been sitting around for a few hours is a different story altogether.) In the rare case where fresh urine has a detectable aroma, it usually takes one of two forms: fruity or eggy. The latter stench most notoriously occurs after eating asparagus and can take pee-ers (and those in the immediate vicinity) by surprise. Pungent Pee has been invoked by some as the sole reason to employ a courtesy flush while going number one.

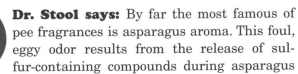 **Dr. Stool says:** By far the most famous of pee fragrances is asparagus aroma. This foul, eggy odor results from the release of sulfur-containing compounds during asparagus digestion. These compounds are absorbed from the digestive tract into the bloodstream and then filtered into urine by the kidneys. Amazingly, asparagus aroma can be detected in urine within fifteen minutes of consuming asparagus! However, it turns out that only 50 percent of the world's population has ever experienced the aroma of asparagus pee. Initially, this was thought to be due to differences in the way some of us digested asparagus, but now it is believed that while all of us excrete these sulfur-containing compounds in our urine, only 50 percent of us have the genes needed to detect their smell. A tip: If you cut off asparagus tips prior to consumption, you won't get Pungent Pee.

CHAPTER 16:

Your Amazing Colon

It's the part of your body where the sun don't shine. But considering all it does, it deserves a little time in the spotlight.

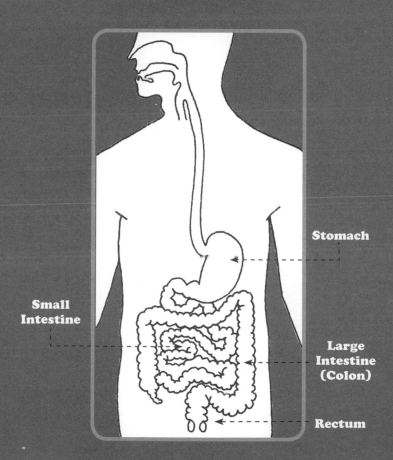

The Long and Windy Road

The GI tract is approximately 25 feet in length. This number has been debated because it depends on how (and when) you measure the length. The intestine is at its longest in a live person and shrinks considerably when it is removed from the body. Despite their names, the large intestine is much shorter than the small intestine, averaging 5 feet in length to the small intestine's impressive 20 feet. However, the large intestine is wider, which is what earned it the name.

Hey Baby, What's Your Time?

The normal gastrointestinal transit time varies from person to person and is also affected by diet. The average individual, however, can expect to see meal remnants (see Déjà Poo, page 81) in their stool 30 to 40 hours after eating. A mere curiosity for most, knowing your transit time may be useful if you are suffering from severe constipation or other gastrointestinal problems such as bloating. Special X-ray tests which track how long it takes radio-labelled food or metallic beads to pass through your GI tract can give an estimate transit time.

●●●●●●●●●●●●●●●● ●●●●●● ●●●●●●●●●●●●●●●● ●●●●●●

Nuggets: *Food moves rapidly through the first 20 feet of small intestine in three to four hours. It's the last five feet of the colon that takes 30-plus hours to traverse.*

Large and Not in Charge

You can survive just fine without a large intestine, and it's not uncommon for it to be partially or even completely removed when it's diseased or injured. Because the large intestine's main job is absorbing water, people who lack one need to drink more, and will also have much wetter stools.

Small But Important

Losing the small intestine is another story. The small intestine is where 90 percent of nutrients are absorbed. When it's diseased, surgeons will try to conserve as much of it as possible, and when it has to be removed entirely, patients can no longer eat. Instead, they'll receive all nutrients intravenously through total parenteral nutrition (TPN). Because they still have a long intestine, they'll still poop.

Borrowed Bowels

Although you don't often hear about them, there is such a thing as a bowel transplant. It's the rarest type of organ transplantation and very complicated to perform, but becoming more common. Bowel transplants are done in cases of acute and chronic intestinal failure, often due to short bowel syndrome. At present, the transplanted intestines come from deceased donors, but researchers are working on ways to graft tissue from living donors in the future.

Stomach Rumblings

Borborygmi is the technical term given to the rumbling noise which emanates from the abdomen. Its physiologic purpose is not to tell us when to eat but rather to sweep the intestines clear of debris left over from the prior meal. Called the "intestinal housekeeper," the MMC, or migrating motor complex, is a massive, wave-like intestinal contraction which begins in the stomach and ends in the large intestine. Occurring every 90 minutes or so during periods of fasting, the MMC rapidly transports gas, fluid and solid debris down the GI tract (it is the gas movement which causes noise). Since the MMC only occurs in the fasting state, the rumbling it produces has come to signify hunger.

Postpartum Poo

Synonyms: *Bonus Baby, Afterbirth, Mother Load*

While many would-be mothers are prepared for the pain of childbirth, few are ready for what could end up being an even more painful delivery—the passing of the first Postpartum Poo. Typically occurring between one and three days after a vaginal delivery (sometimes even longer after a C-section), this bowel movement is a combination of the worst your GI tract has to offer. Some mathematically-inclined mothers have described this experience with the equation:

Postpartum Poo = (Log Jam + Ring of Fire)2.

 Dr. Stool says: Several factors contribute to the pain associated with a woman's first bowel movement after childbirth. First, the lengthy labor process fatigues and stretches the abdominal muscles, thereby making it difficult to generate sufficient expulsive force. Second, the trauma of a baby passing through the vaginal canal puts stress on all nearby organs including the colon, frequently "stunning" it into silence for a couple of hours to a few days. This "stunning" is even more pronounced when the bowel is manipulated during a C-section. Third, women frequently emerge from childbirth extremely dehydrated due to blood loss and the fact that they have been unable to eat or drink for many hours. This creates a hard, desiccated bolus of stool that makes passage more difficult. Fourth, by the end of pregnancy, nearly all women have engorged hemorrhoids. Irritation of these hemorrhoids can make defecation that much more painful. Finally, the presence of an episiotomy or vaginal tear during childbirth can create apprehension toward doing something as simple as sitting, let alone forcing out a rigid bolus of stool.

While the physiologic reality of the postpartum period cannot be changed, several steps may help to make the first few Postpartum Poos more tolerable. Increase water intake to 6 to 8 glasses a day, increase fiber intake, and take frequent walks to help soften the stool and kick-start your GI tract's engine. Gentle laxatives and enemas are also frequently needed in the early postpartum period. The most important thing to do is to go early and go often. Ignoring the urge to poo only makes the bolus harder and drier, making its passage that much more unpleasant.

Sneak Attack

Synonyms: *Ambush Poo, Chocolate Surprise, Deuce Is Loose, Shart*

Regardless of our readiness to "come clean," we have all gambled and lost in this sinister game of shooting craps. We look for a lucky seven, but alas roll snake eyes. It usually starts with the uncomfortable sensations of intestinal rumbling and gaseous bloating. Thinking that a quick, surreptitious release of gas will usher in much needed relief, you prepare for an airy evacuation. But occasionally the anticipated fart contains more than just gas and is accompanied by a liquid smear of poo. In addition to staining your underwear, the smelly remnants of this Sneak Attack will follow you around until you perform the necessary cleaning . . . usually a hasty laundering in the bathroom sink. When faced with a Sneak Attack, we recommend taking care of the cleanup immediately. Often this may require throwing away your underwear and going commando for the rest of the day. A refreshing shower should eliminate all remaining traces of this unwelcome surprise—both physically and emotionally.

 Dr. Stool says: This fecal surprise is due to the presence of liquid stool in the rectum, the antechamber where stool is stored before expulsion. Distension of the rectum (by stool or air) causes the urge to empty the rectum's contents. Normally, solid stool is easily kept inside the vault, but in the setting of significant, watery diarrhea, stool can inadvertently escape when the anal sphincter opens to release gas.

CHAPTER 17:

Throne of Lies

We've been told so many lies about our digestive tract, our poop, and our pee. Here, finally, we expose the truth.

LIE: It takes seven years to digest chewing gum.

While it is true that very few people are walking around with softball-sized, multicolored gumballs rolling around in their stomachs, the admonition against ingesting gum comes from the sound knowledge that the body lacks the ability to digest the rubbery base in chewing gum. Furthermore, this concept of a ball of undigested material in the stomach is called a bezoar and is a real medical entity first described thousands of years ago. A bezoar is simply a collection of undigested material that accumulates in the stomach. Usually seen in patients with a condition called gastroparesis in which the stomach's ability to empty food is severely impaired, a bezoar can also occur in healthy individuals due to consumption of indigestible substances. The most bezoar-genic food? Persimmons.

A particular bezoar known as trichobezoar is formed when individuals consume large quantities of their own hair. This can result in the formation of a hair ball and an honest-to-goodness medical condition called Rapunzel syndrome in which the hair ball causes stomach blockage. As our bodies lack the ability to digest hair, successful removal of this obstructing hair ball is achieved via endoscopic surgery.

Other substances that can form bezoars:
- Apples
- Figs
- Green beans
- Celery
- Sauerkraut
- Antacid medications
- And, yes, bubble gum

Don't fret, there is no need to throw away the rest of the celery and fig salad you were savoring. Consumption of these items (including bubble gum) rarely results in bezoar formation in people whose stomachs are functioning normally. So, unless you have had stomach surgery in the past or have been diagnosed with stomach-emptying problems, don't worry too much about bezoars. But just to play it safe, let's go ahead and hold the persimmons.

LIE: You should wait thirty minutes after you eat before swimming

As kids, we counted the minutes after lunch until we could jump into the pool. On those hot summer days, thirty minutes could seem more like three hundred. Presumably, waiting thirty minutes would allow for adequate digestion of food and prevent vomiting once swimming commenced. Well, it turns out this half-hour respite between eating and swimming has no scientific basis whatsoever.

Studies show that it takes, on average, four hours to completely empty our stomachs. Food needs to be mixed with gastric juices and broken down into smaller particles prior to entering the small intestine (when the risk of vomiting with exertion is thought to be less). The 30-minute moratorium on swimming makes little sense considering that over 90 percent of our hamburger and French fries would still be in the stomach at that point!

Some have argued that waiting 30 minutes after a meal to exercise will help prevent abdominal cramping. Blood diverted to our exercising muscles can theoretically decrease the amount of blood available to our intestinal tract and, thus, cause cramping. There is nothing to support, however, the firm 30-minute moratorium placed on exercise after eating. In fact, there is substantial evidence that light exercise after meals can help with the digestive process.

LIE: Peeing on a jellyfish sting will relieve the pain

Jellyfish cause stinging by injecting an alkali-based fluid through little stingers called nematocysts. Some have claimed that urine's acidity can help neutralize the venom and relieve pain. In fact, normal urine is usually neutral (pH 7) or only slightly acidic and is probably only marginally effective for jellyfish stings. Dr. Stool recommends using a more acidic substance like vinegar (pH 2.5 or so) instead. While it may be inconvenient to carry a bottle of vinegar to the beach, pouring vinegar on your arm is a bit more socially acceptable than peeing on it.

Paradoxical Poo

Synonyms: *Crushing Combo, Excremental Dilemma*

When your constipation gets bad, you can go for days, maybe weeks, without much activity in the number two department. Finally, after much frustration, you may sense that you are on the precipice of a breakthrough and you enter the bathroom preparing to push out the responsible poo plug. However, instead of the free-flowing release of stool, you may be surprised when a spurt of brown liquid splatters mockingly into the toilet. Feeling that the lack of regular defecation has somehow affected your senses, you push again, which results in the expulsion of even more liquid stool. As the days go by, you find only temporary relief from the agony of the logjam through small bursts of unsatisfying, watery diarrhea. While you are glad to be free from total constipation, you are left unfulfilled, wondering when normal, solid stool will return. Is it possible to be constipated and have diarrhea at the same time?

Dr. Stool says: Seemingly flying in the face of poo logic, Paradoxical Poo is the experience of having constipation and diarrhea at the same time. Also called "overflow diarrhea," Paradoxical Poo occurs when watery stool leaks around an unyielding poo plug, a condition called fecal impaction. This ailment is characterized by the presence of a hard bolus of stool which literally plugs up the rectum and prevents passage of normal fecal material. While poo's egress from the body is impeded, liquid stool can seep around the edges and give the appearance of diarrhea, sometimes tricking physicians into prescribing medications that slow down the GI tract and worsen the underlying constipation. Once an accurate diagnosis is made, treatment is with laxatives and, if needed, manual extraction of the poo plug.

The Explosion

Synonyms: *The Bullhorn, Air Show, Sonic Boom, Rolling Thunder*

Rarely heard in public due to societal constraints, the Explosion is characterized by its deafening sound, likened by some to the roar of a lion, by others to the revving of a jumbo jet engine. This deafening expulsion offers particular satisfaction as it usually indicates the release of highly pressurized gas. The decompression can leave you feeling light on your feet, while the raucous sound can bring about hearty congratulations (in the right company, of course).

A fart's loudness is determined by two main factors: the gas's volume and/or pressure, and the aperture through which it passes. Simply put, the greater the gas pressure and/or volume and the smaller the opening of the anal canal, the louder the fart. This association can be expressed mathematically through the following equation:

$$\text{FL (fart loudness)} \approx \frac{\text{Gas}_{\text{volume}} \times \text{Gas}_{\text{expulsion pressure}}}{\text{Anal canal diameter}}$$

Gas volume is determined by two main factors: how much air we swallow and what we eat. All of us unconsciously swallow air as we go about our daily lives. Aerophagia, or excessive air swallowing, typically occurs in people who continuously chew gum or consume a large volume of carbonated beverages. The resultant intestinal buildup of nitrogen-rich gas is typically released by belching, but some of this gas finds its way down the GI tract for a backdoor exit.

A more likely cause of increased gas volumes is the consumption of indigestible carbohydrates found in foods such as chickpeas, cabbage, beans, brussels sprouts, broccoli, onions, and even red wine. These flatogenic foods evade digestion in the small intestine and thus make their way to the colon, where they undergo fermentation by intestinal bacteria. As part of the digestion process, these bacteria let off varying concentrations of carbon dioxide, hydrogen, and methane gas. The precise amount of each gas is dependent on the specific types of bacteria residing in the colon. Contrary to popular belief, only a third of us have the "right" bacteria to produce significant amounts of methane during this digestion process.

Gas expulsion pressure is dependent on the amount of gas to be expelled, but can also be affected by the strength of our abdominal muscle contractions. Abdominal muscle contraction (bearing down) forces stored gas out in a rapid, highly pressurized fashion.

Anal canal diameter: Think of the anal canal as Mother Nature's flute. The sound generated by a fart is dependent on the amount of reverberation it generates while passing through the anal corridor. The smaller the opening, the greater the reverberation and the louder the sound. Some gifted souls have an almost magical control over the size of their anal aperture and can willfully alter the intensity, duration, and pitch of their flatus. Structures like hemorrhoids, which protrude into the anal canal, dial up a fart's volume by rendering flow more turbulent.

Dr. Stool says: Most individuals with explosive flatus suffer from an increased amount of intestinal gas production. Dairy products are a common culprit in those with lactose intolerance. If elimination of milk doesn't do the trick, try avoiding a few common accomplices.

FLATOGENIC FOODS

- Onions
- Coffee
- Lactose
- Apricots
- Beans
- Wheat germ, oats, barley
- Brussels sprouts
- Chickpeas
- Prunes
- Cabbage
- Lentils
- Red wine
- Molasses
- Eggplant
- Broccoli
- Dark beer

After eliminating all dairy and other flatogenic foods from your diet, if your body still produces Explosion farts, try stifling the sound by adopting techniques that will relax the abdomen (deep breathing is one such tactic). Experienced stiflers will actually suck in the abdominal wall in an attempt to decrease the force with which air is expelled. Whatever your tactics, avoid the urge to constrict your anal sphincter: remember, the smaller the opening, the louder the noise. Farting through a tight sphincter can make it seem as though you just tooted on a kazoo. Instead, try slowly releasing the air by gradually relaxing the anal sphincter over a few seconds.

CHAPTER 17:
The Pooture

What will the future bring to the bathroom? Even an expert in scatomancy—the art of reading your future from your poop—can't say for sure, but signs point to a world in which poo becomes ever more valuable. After all, it's a renewable energy source we renew every time we take a dump. Here's what researchers are working on now.

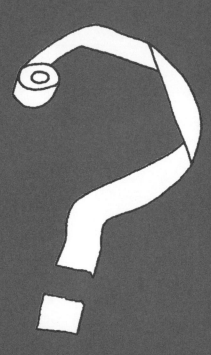

From Poop to Paper

You use paper to clean up your poop. And now, it seems, your poop can be cleaned into paper. An Israeli company called Applied Clean Tech has developed a method to extract cellulose from human waste and other sources of sludge. The poopy cellulose is sterilized, cleaned, and pressed into sheets of paper. The company hopes at some point it will also be transformed into packaging for food and other products. The technology could potentially provide 10 percent of the world's paper supply while reducing landfill sludge waste by 75 percent.

Bike Your Butt Off

Japanese manufacturer Toto is known for its state-of-the-art toilets. Now, they've developed something altogether new: a toilet bike that converts feces directly into biogas. The poop-powered bike can also play music.

Going Brown

Composting toilets are growing in popularity because they reduce household water usage (no flushing involved). Humanure, as this waste is called, decays naturally and can be used as fertilizer (probably better for flowers than your vegetable garden) after about four to six years. Fear not, while it may take years for decamping stool to become enriched soil, poo's atrocious aroma only lasts a week or so.

The Power of Panda Poop

Panda poop has been used to make a number of products, including paper and tea. Now researchers believe it might provide a new energy source. Panda stomachs host uniquely effective microbes that let them extract nutrients from the tough cellulose in bamboo. Researcher Ashli Brown has discovered that these microbes are also very effective at processing lignocellulose, a fiber which holds promise as biofuel material. Nice one, pandas!

Magical Maggots

Farts are gross. Maggots are gross. But maggot farts? Magical! Maggot flatulence has antibiotic properties, which is what makes maggot debridement—using maggots to remove dead tissue—an effective medical treatment when other interventions fail. Researchers are looking into ways to harness this magical healing power that won't require letting maggots eat your flesh.

Going for the Gold

According to a study done recently by the U. S. Geological Survey, the new gold rush may be happening in your colon. The study looked at metal traces in human poop and found surprising amounts of gold, silver, lead, titanium, and zinc. It's estimated that the total worth of the metals flushed away in the U.S. alone may be well over a trillion dollars. As of yet, however, no one's come up with a way to extract it.

Performance-Enhancing Poo

Synonyms: *Anxiety Poo, Preparatory Poo, Running Runs, The Pre-Game Poo*

Sometimes intentional and other times triggered by nerves, the Pre-Game Poo is standard both for competitive athletes and for people with high-pressure presentations looming on the horizon. You never want to have to take a break in the middle of a key proposal or sale, and it would be unheard of to call a time-out for a mid-game bathroom break. With empty bowels, you can run faster and jump higher. Similarly, the absence of stool in your colon will make your presentation crisper and diminish fears of an unscheduled pit stop or a loud gaseous emission. Unplanned Performance-Enhancing Poos often take on a more liquid consistency than their planned counterparts. Although these poos may not have the grandeur of some, their timing is critically important.

PERFORMANCE-ENHANCING PEE

While some baseball players fear the discussion of pee, thinking that a drug test is soon to follow, there are some who harness the power of pee to boost on-field performance. Several major leaguers, including Moises Alou and Jorge Posada, forego batting gloves and pee on their hands to prevent and relieve calluses. Similarly, heavyweight boxing champ Vitali Klitschko uses an old Ukranian secret passed down to him by his grandmother to keep his hands from swelling up like other boxers' hands. Yes, he wraps his hands in his infant's diapers before a match.

These athletes have figured out what urine therapy believers have been saying for centuries—massaging your skin with urine can keep it healthy and soft. The dermatologic benefits of urine are largely due to urea, a substance that can be also be found in many commercially available lotions. Although urine is sterile, Dr. Stool feels obligated to recommend that batboys wear gloves at all times and that you avoid shaking hands with Klitschko.

Curtain Call

Synonyms: *Overtime, Groundhog Day, The Remix*

Sometimes you finish defecating, wipe, pull up your pants, flush the toilet, and suddenly feel a stomach grumble. Despite the fact that you thought you were done with your business, your GI tract is ostensibly ringing the bell for Round 2, and you need to get back in the ring. If you don't answer the call, things could get ugly, fast.

Dr. Stool says: Yet another instance in which the sequel turns out not to be as good as the original. This poo most often occurs as a result of a normal physiologic process termed the MMC, or migrating motor complex. Occurring at 90-minute intervals, the MMC is a massive sweeping motion of the colon that quickly propels stool downstream (picture a huge tidal wave crashing ashore). If one of these waves happens to "refill" the rectum after you thought you finished the deed, there's nothing to do but unleash the second edition. On the bright side, at least the toilet seat will be warm.

However, the feeling of constantly needing to empty one's bowels, known as tenesmus, could be a sign of a serious underlying medical condition. When accompanied by rectal pain and bleeding, this most unpleasant sensation can be the first sign of an intestinal inflammation, most commonly a result of either ulcerative colitis or Crohn's disease.